Other Monographs in this Series

AMS (MOS) subject classifications (1970).

Primary 20D20, 20A02; Secondary 20D05.

Library of Congress Cataloging in Publication Data

Glauberman, G   1941-
   Factorizations in local subgroups of finite
groups.

   (Regional conference series in mathematics ;
no. 33)
   "Based on lectures given at a conference in
Duluth in 1976."
   Bibliography: p.
   Includes index.
   1. Finite groups. 2. Sylow subgroups.
I. Title. II. Series.
QA1.R33 no. 33 [QA171]   510'.8s [512'.2]
ISBN 0-8218-1683-7   77-13373

Conference Board of the Mathematical Sciences

CBMS

Regional Conference Series in Mathematics

Number 33

Factorizations in Local Subgroups of Finite Groups

G. Glauberman

Published for the
Conference Board of the Mathematical Sciences
by the
American Mathematical Society
Providence, Rhode Island
with support from the
National Science Foundation

Contents

To Gertrude and William Zwillinger

Preface

The past two decades have seen an extraordinary flowering of finite group theory. Most of this work has been aimed toward the specific problem of determining the finite simple groups. However, this pursuit has stimulated the study of other aspects of finite groups which are of interest in themselves as well as in their applications to simple groups. The purpose of this monograph is to describe some recent progress in one such aspect, that of Sylow subgroups. In particular, we address ourselves to the following question:

Given a prime p and a Sylow p-subgroup S of a finite group G, how is the structure of G influenced by the structure of S and the manner in which S is embedded in G?

We will focus on two related special cases of this question:

Question 1. Which elements of S are conjugate in G?

Question 2. What is the relation between S and G if $C(O_p(G)) \subseteq O_p(G)$?

The first important connections between S and G to be proved were obtained by Burnside and Frobenius about the turn of the century. However, their results applied only to special situations. The true strength of the connections between S and G in the general situation was first suggested by the techniques of John Thompson's Ph. D. dissertation in 1959. These techniques were developed and applied in the Odd Order Paper, Thompson's N-group paper, and many later articles. At a conference in Oxford in 1969, I gave a report on this subject entitled *Global and Local Properties of Finite Groups* (abbreviated here by GL); the proceedings of the conference appeared in the book, *Finite Simple Groups*, edited by M. B. Powell and G. Higman. The present monograph is based on lectures given at a conference in Duluth in 1976 and is intended to supplement GL by describing some of the progress since that time.

The text of GL was divided into two parts. The first half consisted of results by Alperin, Gorenstein, Thompson, and others, which developed a theoretical framework for investigating Question 1 and reducing it to Question 2. In the second half of GL, the methods of Thompson, Wielandt, and others were used to prove some special cases of Question 2 for odd p and thus obtain partial answers to Question 1.

The main part of this book follows the same pattern as GL, but with a different emphasis. The major goal in GL was to derive as much information as possible from the normalizer $N(W)$ of some characteristic subgroup W of S; this was done by reducing to special cases of Question 2 and then showing that $N(W) = G$. In contrast, the major goal in this book is to derive information from the normalizers of a *set* of characteristic subgroups by proving factorizations of the form $G = N(W_1)N(W_2)$. The reason for this is that recent results cover different territory (e.g., $p = 2$) from that in GL, and it has been possible (so far) to prove only factorizations, but not normality, for these results.

The body of this book is divided into three chapters as follows:

Chapter I. Here we prove some new results which extend the reductions of Question 1 in GL to cover an arbitrary set of characteristic subgroups rather than a single characteristic subgroup. To do this, we first review some concepts and results from the first half of GL. We end the chapter in a quite different vein with some important new work of Yoshida on transfer.

Chapter II. The second half of GL examined Question 2 in detail for p odd. In particular, one result (similar to the author's "ZJ-Theorem") gave a sufficient condition to have $N(W) = G$ for a specific nonidentity characteristic subgroup W of S. Unfortunately, the proof collapses completely for $p = 2$. Recently, a weak analogue for $p = 2$ has been obtained by using factorizations, and most of Chapter II is devoted to proving this result. In the last section of Chapter II, we apply this result and an important theorem of Goldschmidt to classify the simple groups in which the symmetric group of degree four is not involved.

Chapter III. Here we remove most of our earlier restrictions on G; in particular, we allow $SL(2, p)$ to be involved in G. In this situation, practically none of our previous machinery is applicable. We discuss without proofs a variety of new results and conclude with some applications, results in related areas, and open questions.

Having described what is contained in this book, we must describe what has been omitted. Fortunately, very little in the way of proofs. The topic of Sylow subgroups studied here differs from some other topics in finite group theory in not requiring an extensive background and long proofs. Indeed, it was only with extreme reluctance that we refrained from proving two theorems of Sylow at the beginning of GL. Except for these theorems and a few other quoted results, the main part of GL is self-contained. The present work assumes somewhat greater familiarity with elementary group theory but is otherwise largely self-contained (except for Chapter III, as mentioned above). There are three major exceptions.

(1) In Chapter I, many results are quoted from GL. (Since most of the results and techniques of the present work complement rather than generalize those of GL, we recommend but do not assume familiarity with GL.)

(2) Most of Chapter II is devoted to the proof of a single result (Theorem B). This result was first proved only for solvable groups. However, for the sake of applications to simple groups, the result was extended to allow composition factors of the form $PSL(2, 3^{2n+1})$ and $Sz(2^{2n+1})$. As we are interested only in using the latter groups rather than studying them, the reader unfamiliar with them is welcome merely to assume all the information about them needed in the proof of Theorem B ($\S\S1-5$) or simply to assume throughout that G is solvable. The same remarks apply to a preliminary result, Theorem A, which is proved in Appendix A1.

(3) The applications to simple groups and other topics at the end of Chapter II require several important facts which must be quoted without proof.

Unfortunately, there are many topics closely related to ours which space does not permit us to include. Our initial question asks how S influences the structure of G. In this book, we restrict our attention mainly to the influence of S on the internal, or 'local' structure of G, as exemplified by Questions 1 and 2. However, there is much research directed toward the 'global' structure of G, i.e., the structure of G as a whole. In particular, there are many results which yield the precise structure of G when $p = 2$, G is simple, and S (or the centralizer of an element of order two in S) is isomorphic to a given group. Unfortunately, most work of this type depends on such 'global' tools as ordinary character theory, block theory, or generators and relations. In comparison with the 'local' theory, papers in this area generally require more breadth, more depth, and more length. For this reason, we give only a brief sample of this work in the section on simple groups in Chapter II and in the quoted results in Chapter III.

Another large, related question that we have been forced to omit is how the structure of G is influenced by restrictions on its Sylow subgroups for more than one prime. In other words, how do the p-subgroups and q-subgroups of G interact when $q \neq p$, and how can we integrate information obtained for individual primes? Answers to this question have produced many applications to signalizer functors, uniqueness theorems for simple groups, and fixed-point-free groups of automorphisms. Fortunately, the reader can obtain an excellent introduction to this subject from three short papers [1], [11], [12] recommended as preliminary reading for participants in the conference. In particular, [1] gives an expository account of the entire area of relations between local and global information.

We emphasize one important convention which we will observe. Throughout the book, G will denote a fixed but arbitrary finite group; p, a fixed but arbitrary prime; and S, a fixed but arbitrary Sylow p-subgroup of G. Of course, all of the groups considered in this book are finite.

It is a great pleasure for me to thank the institutions which made the Duluth Conference and this book possible—the Conference Board of the Mathematical Sciences, the University of Minnesota at Duluth, and the National Science Foundation (which also assisted by research grants). I have benefited very much from the comments of the participants, especially I. M. Isaacs, who also contributed the material in §1.6 of this book. In addition, I am indebted to R. Niles and T. Yoshida for permission to use their unpublished work, and to Chat-Yin Ho for many corrections to GL which appear in the Appendix. Most of all, I wish to thank Joseph Gallian for a year of tireless effort to arrange, organize, and run the Conference superlatively.

I would also like to acknowledge my gratitude to L. A. Shepp and George Bachman, who began my education in group theory, and to J. G. Thompson and H. Wielandt for their further encouragement and stimulation. I am particularly grateful to R. H. Bruck as a teacher and friend for many years. In a sense, Daniel Gorenstein and Richard Lyons are the real authors of this book, because nearly all of its contents have grown out of two problems which they proposed two years ago.

Finally, I wish to express my thanks to Mr. Fred Flowers for an excellent job of typing under extremely trying circumstances and to my wife (and copy editor) and my children, Daniel and Rachel, for coping with similar circumstances.

University of Chicago   George Glauberman

Preface to Second Printing

As mentioned above, this work supplements the author's earlier report, GL, on the connection between global and local properties of finite groups. Since the first printing, there has been significant progress in this area, and we have made corresponding additions to the text:

(1) Bernd Stellmacher has solved an outstanding open problem (Question 4.1 in Section 3.4) concerning an analogue to K_∞ and ZJ for $p = 2$. This yields a simpler proof of the most important result of Chapter II, the classification of non-abelian simple groups in which the symmetric group of degree four is not involved (Theorem C). His proof uses Section II.2 and Appendix A1 of this work. We have added a new section, Appendix A3, which is an account by the author of Stellmacher's article, written for Mathematical Reviews, with the kind permission of the American Mathematical Society.

(2) Several authors have solved other open questions in GL and Section 3.4. We have extended Section 3.4 to mention these solutions.

(3) There has been an explosion of activity in the area of Sections 3.1 and 3.2, "pushing-up" (including "failure of factorization"), especially after a landmark article by David Goldschmidt on amalgams in 1980. Space permits us only to mention here an expository introduction to new ideas in this subject in Sections 4.12 and 4.13 of Daniel Gorenstein's book [73], published in 1982.

In addition, all known errors have been corrected.

I thank Saunders Mac Lane for valuable advice on preparing this printing and Andrew Chermak and Ronald Solomon for information about current research.

University of Chicago   **George Glauberman**

Chapter I. Reductions to Local Subgroups and Sections

1. Introduction. We mentioned the following two problems in the Preface:

Question 1. Which elements of S are conjugate in G?

Question 2. What is the relation between S and G if $C(O_p(G)) \subseteq O_p(G)$?

In this chapter, we address ourselves to Question 1 and show that in many instances it may be reduced to Question 2.

Two elements of S are said to be *fused* in G if they are conjugate in G. Thus, fusion in S defines an equivalence relation on S. In general, the normalizer in G of a nonidentity group H of prime-power order is called a *local* subgroup of G; it is called a *p-local* subgroup of G if H is a p-group. A subgroup K of G is a *normal p-complement* in G if $G = KT$ and $K \cap T = 1$ for every Sylow p-subgroup T of G.

Let us consider some results on fusion taken from GL. To avoid triviality, we will assume that $S \neq 1$.

The first general answer to Question 1 was obtained by Burnside:

If S is Abelian, then two elements of S are fused in G if and only if they are fused in $N(S)$.

Burnside found a powerful application of his answer:

If S is contained in the center of its normalizer, then G has a normal p-complement.

Suppose we try to follow in Burnside's footsteps. In his case, the fusion of S in G is determined completely by the single p-local subgroup $N(S)$. But what if S is not Abelian? Obviously, two elements of S that are fused in $N(S)$ must be fused in G, but the converse need not hold. Consider, however, the family of *all* p-local subgroups of G. A surprising result of Alperin (Theorem 3.1) shows that this family completely determines the fusion of S in G. This is what makes fusion truly a local property of G.

Perhaps equally surprising, but proved much earlier than Alperin's Theorem, is the fact that the fusion of S in G determines whether G has any nontrivial factor groups which are p-groups. This follows from a result known as the Focal Subgroup Theorem (Theorem 3.4), which is proved by applying the transfer homomorphism. Burnside's normal p-complement result is a consequence of this result (Theorem 3.6(b)). (Burnside's fusion result follows from Lemma 4.5 of GL.) We may paraphrase Alperin's Theorem and this result by saying that *local subgroups determine fusion, and fusion determines transfer.*

Now let us take a second look at Burnside's fusion result. In it, fusion is determined by the single p-local subgroup $N(S)$. Alperin's Theorem tells us that, in general, fusion is determined by the family of *all* p-local subgroups of G. It turns out that in many groups, fusion is determined by one p-local subgroup, possibly $N(S)$ or possibly a larger group (e.g., the normalizer of a characteristic subgroup of S). It was proved by Alperin and Gorenstein that such a subgroup will exist if an analogous subgroup exists within each of the p-local subgroups of G. Further investigation shows that it will exist if an analogous subgroup exists within each of the sections G^* of G for which $C_{G^*}(O_p(G^*)) \subseteq O_p(G^*)$. (Here, the concept of *p-stability* introduced by Gorenstein and Walter becomes relevant if p is odd.) This situation is described as "control of strong fusion" in GL; one may similarly investigate "control of transfer".

We have now described the principal ideas of §§3–6 of GL; they may be summarized by section numbers as follows:

3. Local subgroups determine fusion.

4. Fusion determines transfer.

5. Local control yields global control.

6. Control of sections yields global control.

In this chapter, we will review the main results of these sections of GL and prove some related results. In particular, some results of GL about control by a single p-local subgroup will be generalized to results about joint control by a given set of p-local subgroups. We will conclude with some exciting new work of Yoshida on transfer.

2. Notation and preliminary results. Most of our notation is standard and agrees with GL. However, we note the following expressions which are not entirely standard:

$H \subseteq G$, $G \supseteq H$: H is a subgroup of G;

$H \subset G$, $G \supset H$: $H \subseteq G$ and $H \neq G$;

$H \not\subseteq K$, $K \not\supseteq H$: H is not contained in K (for H, $K \subseteq G$);

$[H, K, L]$: $[[H, K], L]$ (for H, K, $L \subseteq G$);

$A - B$: $\{x \in A | x \notin B\}$ (for subsets A, B of G).

In addition, we differ from GL and agree with [37] by writing $H \triangleleft G$ and $G \triangleright H$ to mean that H is a normal subgroup of G (denoted by $H \trianglelefteq G$ and $G \trianglerighteq H$ in GL).

We also require some further notation, most of which is not given in GL. If G is an operator group on a group V (GL, p. 51) and $H \subseteq G$, then

$$C_V(H) = \{v \in V | v^h = v \text{ for all } h \in H\}.$$

We say that H *centralizes* V if $C_V(H) = V$. For $g \in G$, we define $C_V(g) = C_V(\langle g \rangle)$ and say that g *centralizes* V if $\langle g \rangle$ does. A subgroup W of V is *fixed* by G if every element of G maps W to itself. Let Aut G and In G denote the group of all automorphisms and the group of all inner automorphisms of G respectively.

As is usual, let $O_p(G)$ and $O_{p'}(G)$ denote the largest normal p-subgroup and the largest normal p'-subgroup of G. Then $O_{p,p'}(G)$ is defined by

$$O_{p,p'}(G)/O_p(G) = O_{p'}(G/O_p(G)).$$

Let $O^p(G) = \langle x | x \text{ is a } p'\text{-element of } G \rangle$ and $O^{p'}(G) = \langle x | x \text{ is a } p\text{-element of } G \rangle$. If G is a p-group, let

$$\Omega_1(G) = \langle x \in G \mid x^p = 1 \rangle \quad \text{and} \quad \Omega_1 Z(G) = \Omega_1(Z(G)).$$

We use Z_p to denote both the field of the integers, modulo p, and its additive group.

Again assume that G is arbitrary and not necessarily a p-group. Let $\Phi(G)$ be the *Frattini subgroup* of G, i.e., the intersection of the maximal subgroups of G. A *section* of G is a factor group H/K for which $K \subseteq H \subseteq G$ and $K \lhd H$. A group G^* is *involved* in G if it is isomorphic to some section of G. A section H/K of G is a *chief factor* of G if $H, K \lhd G$, and $H/K \neq 1$, and if no normal subgroup N of G satisfies the condition $K \subset N \subset H$. In this case, G operates on H/K by conjugation; H/K is said to be a *central* chief factor of G if H/K is centralized by G, and is said to be a *noncentral* chief factor of G otherwise. For $L \lhd G$, we say that a chief factor H/K of G is *within* L if $H \subseteq L$.

As in GL, we say that G is a *p-stable* group if it satisfies the following condition:

Whenever P is a p-subgroup of G, $x \in N(P)$, and $[[P, x], x] = 1$, then the coset $xC(P)$ lies in $O_p(N(P)/C(P))$.

We showed in GL (Proposition 7.8) that this is the same definition as that of [37].

An element of G is called an *involution* if it has order two. If $x, y \in G$, then x *inverts* y if $y^x = y^{-1}$. A nonempty subset A of G is a *trivial intersection set* in G if it satisfies the following condition:

For every $x \in G$, $A^x = A$ or $A^x \cap A$ is a subset of $\{1\}$.

Suppose n is an arbitrary natural number. We let S^n and A^n denote the symmetric and alternating groups of degree n. For every prime power q, let $SL(n, q)$ denote the *special linear group* of degree n over the field $GF(q)$ with q elements, i.e., the group of all matrices of degree n and determinant 1 over $GF(q)$. Then we let

$$PSL(2, q) = SL(2, q)/Z(SL(2, q)).$$

It is well known [44, p. 182] that $PSL(2, q)$ is simple except when $q = 2$ or $q = 3$.

For any finite-dimensional space V over a field F, let $SL(V, F)$ be the *special linear group* of V over F, i.e., the group of all linear transformations of V over F of determinant 1.

Finally, we denote the groups in Suzuki's family of simple groups by $Sz(q)$ for $q = 2^{2n+1}$ and all natural numbers n. They are described in pp. 133–143 of [57].

Lemma 2.1. *The groups $O_p(G)$ and $O_{p'}(G)$ centralize each other.*

Proof. Since $O_p(G), O_{p'}(G) \lhd G$,

$$[O_p(G), O_{p'}(G)] \subseteq O_p(G) \cap O_{p'}(G) = 1.$$

Lemma 2.2. *Suppose $N \lhd G$. Then*

(a) *$N \cap S$ is a Sylow p-subgroup of N,*

(b) *SN/N is a Sylow p-subgroup of G/N,*

(c) *$G = SN$ if and only if G/N is a p-group,*

(d) *$O_p(G/N) \subseteq SN/N$,*

(e) *for every subgroup R of S, the group $RN \cap S$ is a Sylow p-subgroup of RN,* and

(f) *if H/K is a chief factor of G and H/K is an elementary Abelian p-group, then $O_p(G/C_G(H/K)) = 1$ and $O_p(G) \subseteq C(H/K)$.*

Proof. Parts (a), (b), (c) form Lemma 2.10 of GL. Part (d) follows from (a) and (c). In (e), $|N \cap S| = |N|_p$ by (a); then

$$|RN \cap S| \geq |R(N \cap S)| = |R/(R \cap N \cap S)||N \cap S|$$
$$= |R/(R \cap N)||N \cap S|$$
$$= |RN/N||N|_p = |RN/N|_p|N|_p = |RN|_p.$$

Part (f) follows from Proposition 7.6 of GL.

Lemma 2.3 (Frattini argument). *Suppose $H \lhd G$ and T is a Sylow p-subgroup of H. Then $G = N(T)H = HN(T)$.*

Proof. This is Lemma 2.11 of GL.

Lemma 2.4. *Let X be a group. Suppose that X is involved in a section H/K of G for which $O_p(H/K) \neq 1$. Then there exists a subgroup L of H for which X is involved in L and*

$$H = LK \quad and \quad 1 \subset O_p(H/K) = O_p(L)K/K.$$

Proof. This is a slight extension of Lemma 10.6 of GL and follows from its proof.

Lemma 2.5. (a) *If $H \subseteq G$ and $G = \langle H, \Phi(G) \rangle$, then $G = H$.*
(b) *If $S_1, \ldots, S_n$ are elements or subsets of G and $G = \langle S_1, \ldots, S_n, \Phi(G) \rangle$, then $G = \langle S_1, \ldots, S_n \rangle$.*
(c) *If $N \lhd G$ and $N \subseteq \Phi(G)$, then $\Phi(G/N) = \Phi(G)/N$.*

Proof. (a) Here, if $H \subset G$, then H is contained in a maximal subgroup M of G. By the definition of $\Phi(G)$, we obtain that $M \supseteq \Phi(G)$ and $M \supseteq \langle H, \Phi(G) \rangle = G$, a contradiction.

(b) Let $H = \langle S_1, \ldots, S_n \rangle$ in (a).

(c) Every maximal subgroup of G/N has the form M/N for a maximal subgroup M of G. Since $N \subseteq \Phi(G)$, every maximal subgroup M of G arises in this way. Hence

$$\Phi(G/N) = \bigcap_{M \text{ maximal in } G} M/N = \left(\bigcap_{M \text{ maximal in } G} M \right) \Big/ N = \Phi(G)/N.$$

Lemma 2.6. *Suppose T is a p-group. Then*
(a) *$T/\Phi(T)$ is elementary Abelian,*
(b) *$\Phi(T) = 1$ if and only if T is elementary Abelian,*
(c) *$\Phi(T) = \langle T', x^p | x \in T \rangle$, and*
(d) *(Burnside) every p'-subgroup of $\mathrm{Aut}\, T$ operates faithfully on $T/\Phi(T)$.*

Proof. Parts (a), (b), and (d) are given on page 174 of [37]. Let $T^* = \langle T', x^p | x \in T \rangle$. By (a), $\Phi(T) \supseteq T^*$. Now (c) follows from Lemma 2.5 (c).

Lemma 2.7. *Suppose T is an Abelian p-group and G is a p'-group of automorphisms of T. Then*

(a) $T = [T, G] \times C_T(G)$;

(b) G operates faithfully on $\Omega_1(T)$.

Proof. (a) This is Theorem 5.2.3, page 177 of [37].

(b) For each $g \in G$, part (a) yields that

$$1 = \Omega_1([T, \langle g \rangle]) \cap C_T(\langle g \rangle) \subset \Omega_1([T, \langle g \rangle]).$$

Lemma 2.8. *Suppose that T is a finite p-group and $T = T_0 \supseteq T_1 \supseteq \cdots \supseteq T_n = 1$ is a series of normal subgroups of T. Let A be the group of all automorphisms of T that fix $T_0, T_1, \ldots, T_n$. Let B be the subgroup of A consisting of the elements of A that fix every coset of T_{i-1} in T_i ($i = 1, \ldots, n$). Then B is a normal p-subgroup of A.*

Proof. This is Lemma 10.4 of GL.

Definition. In the above situation, we say that every element of B *stabilizes* the chain $T = T_0 \supseteq T_1 \supseteq \cdots \supseteq T_n = 1$. More generally, any operator on T that yields an automorphism in B is said to *stabilize* the above chain.

Lemma 2.9. *Suppose $H, K \lhd G$ and $K \subseteq H$ and H/K is a p-group. Assume that every chief factor of G within H/K is central. Then $O^p(G)$ centralizes H/K.*

In particular, if H is a normal p-subgroup of G and every chief factor of G within H is central, then $O^p(G)$ centralizes H.

Proof. Take a chief series of G that passes through H and K:

$$G = G_0 \supseteq G_1 \supseteq \cdots \supseteq G_r = H \supseteq G_{r+1} \supseteq \cdots \supseteq G_s = K \supseteq \cdots \supseteq 1.$$

Then apply Lemma 2.8 to the normal series of H/K given by

$$H/K = G_r/K \supseteq \cdots \supseteq G_s/K = 1.$$

Lemma 2.10 (Baer, Suzuki, J. H. Walter, Alperin, Lyons, Wielandt). *Suppose $x \in G$ and, for every $g \in G$, the group $\langle x, x^g \rangle$ is a p-group. Then $x \in O_p(G)$.*

Proof. There are many proofs of this result, which was originally proved by R. Baer. Among the most interesting are those of Alperin and Lyons [23, p. 5] and H. Wielandt [66]. Some others are given in pages 3–5 of [23].

3. Definitions and basic properties. In this section, we review some of the concepts and results of §§3 and 4 of GL.

Definitions. Let $\mathcal{F}$ be a set of subgroups of S. Suppose that A and B are nonempty subsets of S and $g \in G$. We say that A is $\mathcal{F}$-conjugate to B via g if there exist subgroups $T_1, \ldots, T_n$ in $\mathcal{F}$ and elements $g_1, \ldots, g_n$ of G such that:

$g_i \in N(T_i)$ $(i = 1, \ldots, n)$, $\langle A \rangle \subseteq T_1$ and $\langle A \rangle^{g_1 \cdots g_i} \subseteq T_{i+1}$ $(i = 1, \ldots, n-1)$, and $A^g = B$ and $g = g_1 \cdots g_n$.

We say that $\mathcal{F}$ is a *conjugation family* (*for S in G*) if it has the following property: Whenever A and B are nonempty subsets of S and $g \in G$ and $A^g = B$, then A is $\mathcal{F}$-conjugate to B via g.

Let $\mathcal{S}(G)$ be the set of all sequences $x = (x_1, \ldots, x_n)$ of distinct elements of G. For each such sequence x and each $g \in G$, let $x^g = (x_1^g, \ldots, x_n^g)$. We say that two elements x, y of $\mathcal{S}(G)$ are *conjugate* in G if $y = x^g$ for some $g \in G$.

Suppose that $\sim$ is an equivalence relation on S (or on $\mathcal{S}(S)$). We will write

$x \sim y$ to mean $(x, y) \in \sim$,

$x \not\sim y$ to mean $(x, y) \notin \sim$.

Let $H \subseteq G$. We say that $\sim$ *contains fusion* in H if $x \sim y$ whenever $x, y \in S \cap H$ (or $x, y \in \mathcal{S}(S \cap H)$) and x and y are conjugate in H.

Note that $\sim$ contains fusion if and only if $\sim$ contains (x, y) for every pair x, y of elements (or sequences) in $S \cap H$ that are conjugate in H. This explains the phrase, "contains fusion".

The first major step in our subject was Alperin's proof that conjugacy functors exist. In particular, it tells us that two elements or subsets of S are conjugate in G if and only if they are 'locally conjugate' in G.

Theorem 3.1 (Alperin). *Let $\mathcal{F}$ be the set of all subgroups T of S for which $N_S(T)$ is a Sylow p-subgroup of $N_G(T)$. Then $\mathcal{F}$ is a conjugation family for S in G.*

This result is Theorem 3.5 of GL. We shall also require the following results, which are Lemma 3.6 and Proposition 3.7 of GL.

Lemma 3.2. *Suppose that $\mathcal{F}$ is a conjugation family for S in G. Let $\mathcal{F}'$ be the set of all $T \in \mathcal{F}$ that contain $O_p(G)$. Then $\mathcal{F}'$ is a conjugation family for S in G.*

Proposition 3.3. *Let $\sim$ be an equivalence relation on S (or on $\mathcal{S}(S)$) and let $\mathcal{F}$ be a conjugation family for S in G. Assume that $\sim$ does not contain fusion in G. Then there exist $T \in \mathcal{F}$ and $x, y \in T$ (or $x, y \in \mathcal{S}(T)$) such that x and y are conjugate in $N(T)$, $x \not\sim y$, and $T \supseteq O_p(G)$.*

As far as we know, there is only one kind of nonsimplicity criterion for G that is determined completely by local information and is independent of the prime p. This is the criterion of having a nontrivial Abelian factor group. We see this from the following result, which is Theorem 4.1 of GL (proved in pages 245–251 of [37]).

Theorem 3.4 (Focal Subgroup Theorem). *We have*

$$S \cap G' = \langle x^{-1}y \,|\, x, y \in S \text{ and } x \text{ is conjugate to } y \text{ in } G \rangle.$$

We shall require the following related results (Proposition 4.4, Theorem 4.8, and Proposition 4.9 of GL). Note that the first seems to be difficult to obtain from fusion information alone, e.g., from the Focal Subgroup Theorem; its proof in GL uses the transfer map.

Proposition 3.5. *We have* $S \cap Z(G) \cap G' = S \cap Z(G)$.

Theorem 3.6 (Burnside). (a) *If S is Abelian, then* $S \cap G' = [S, N(S)]$.
(b) *If* $S \subseteq Z(N(S))$, *then G has a normal p-complement.*

Proposition 3.7. *Let* $S^* \subseteq S$. *Let $\sim$ be the equivalence relation on S given by* $x \sim y$ *if* $S^* x = S^* y$. *Suppose that* $H \subseteq G$, $T \subseteq S$, *and T is a Sylow p-subgroup of H. Then $\sim$ contains fusion in H if and only if* $H' \cap T \subseteq S^*$.

The next result follows from Lemma 6.4 of GL and Lemma 2.1. (By Lemma 2.1, the group H in Lemma 6.4 of GL must be $O_{p'}(G)$.)

Lemma 3.8. *Let* $T = O_p(G)$. *Suppose that* $C_S(T) \subseteq T$. *Then*
(a) $C(T) = Z(T) \times O_{p'}(G)$, *and*
(b) *for* $\overline{G} = G/O_{p'}(G)$, $C_{\overline{G}}(O_p(\overline{G})) \subseteq O_p(\overline{G})$.

4. Sets of conjugacy functors. In §5 of GL, we proved some results of Alperin and Gorenstein about conjugacy functors and well-placed subgroups. This work has been extended by D. Finkel to groups with certain factorizations of subgroups. In this section, we obtain Finkel's results and some generalizations by extending various results in GL about a single conjugacy functor to results about sets of conjugacy functors.

Definitions. Let $\mathcal{C}_p(G)$ be the set of all p-subgroups of G. A *conjugacy functor* (for the prime p) on G is a mapping W of $\mathcal{C}_p(G)$ into $\mathcal{C}_p(G)$ that satisfies the following three conditions for every $P \in \mathcal{C}_p(G)$:
(i) $W(P) \subseteq P$;
(ii) $W(P) \neq 1$ if $P \neq 1$; and
(iii) $W(P^g) = (W(P))^g$ for every $g \in G$.

We shall not need the actual definition of well-placed subgroups of S (with respect to W, S, and G), but only the properties in the following result (which actually characterize the class of well-placed subgroups).

Lemma 4.1. *Suppose W is a conjugacy functor on G and* $T \subseteq S$.
(a) *There exists a well-placed subgroup of S that is conjugate to T in G.*
(b) *If T is a well-placed subgroup of S, then $N_S(T)$ is a Sylow p-subgroup of $N_G(T)$ and $W(N_S(T))$ is a well-placed subgroup of S.*

Proof. These are Lemma 5.2 and Lemma 5.1 (a) of GL.
The next result is Theorem 5.3 of GL.

Theorem 4.2 (Alperin-Gorenstein). *Let W be a conjugacy functor on G and let $\mathcal{F}$ be the set of all well-placed subgroups of S with respect to W. Then $\mathcal{F}$ is a conjugation family for S in G.*

The following result generalizes Proposition 5.4 of GL from a single conjugacy functor to a set of conjugacy functors. All of our further generalizations depend upon it.

Theorem 4.3. *Suppose that $\sim$ is an equivalence relation on S or on $\mathcal{S}(S)$, $\mathcal{W}$ is a set of conjugacy functors on G, and $\mathcal{F}$ is the set of all subgroups T of S for which*

$N_S(T)$ *is a Sylow* p-*subgroup of* $N(T)$. *Assume that*

(i) *for every* $W \in \mathbb{W}$, $\sim$ *contains fusion in* $N(W(S))$, *and*

(ii) $\sim$ *does not contain fusion in* G.

Then there exists some $T \in \mathcal{F}$ *that satisfies the following conditions:*

(a) *for each* $W \in \mathbb{W}$, $\sim$ *contains fusion in* $N(W(N_S(T)))$ *and in* $N(W(N_S(T))) \cap N(T)$,

(b) $\sim$ *does not contain fusion in* $N(T)$, *and*

(c) *either*

(ci) $T \supseteq O_p(G)$ *and there exist* $x, y \in T$ *(or* $x, y \in \mathcal{S}(T)$*) for which* x *is conjugate to* y *in* $N(T)$ *and* $x \not\sim y$, *or*

(cii) *there exist* $Q \in \mathcal{F}$, $W \in \mathbb{W}$, *and a subgroup* R *of* S *that properly contains* $O_p(G)$ *such that* $T = W(N_S(Q))$, $\sim$ *does not contain fusion in* $N(Q)$, *and* Q *is equal to* R *or to* $W'(R)$ *for some* $W' \in \mathbb{W}$.

Proof. Let $P = O_p(G)$. By Theorem 3.1 and Lemma 3.2,

(4.1) the elements of $\mathcal{F}$ that contain P form a conjugation family for S in G.

By Proposition 3.3, there exist some $T_0 \in \mathcal{F}$ and $x_0, y_0 \in T_0$ (or $x_0, y_0 \in \mathcal{S}(T_0)$) such that $T_0 \supseteq P$, x_0 and y_0 are conjugate in $N(T_0)$, and $x_0 \not\sim y_0$. If $T_0 = P$, we may let $T = T_0$.

Assume that $T_0 \supset P$. Let $\mathcal{F}'$ be the set of all $T \in \mathcal{F}$ that satisfy (b) and (c). Then $T_0 \in \mathcal{F}'$, whence $\mathcal{F}'$ is not empty. Choose $T \in \mathcal{F}'$ such that $|N_S(T)|$ is maximal, that is,

(4.2) $$|N_S(T)| \geq |N_S(T^*)| \quad \text{for all} \quad T^* \in \mathcal{F}'.$$

To complete the proof it will suffice to verify (a).

Suppose $W \in \mathbb{W}$ and (a) is violated for W. Then

(4.3) $$\sim \text{ does not contain fusion in } N(W(N_S(T))).$$

We will use (4.3) to derive a contradiction. By condition (i) of the hypothesis, $N_S(T) \neq S$. By (c), $P \subset N_S(T)$. Hence

(4.4) $$P \subset N_S(T) \subset S.$$

Let S^* be an arbitrary element of $\mathcal{F}$ for which $S^* \supseteq P$ and $|S^*| \geq |N_S(T)|$. Then $S^* = S$ or $N_S(S^*) \supset S^*$. By (4.4),

(4.5) $$|N_S(T)| < |N_S(S^*)|.$$

Take any $x, y \in S^*$ (or $x, y \in \mathcal{S}(S^*)$) such that x and y are conjugate in $N(S^*)$. If $x \not\sim y$, then (b) and (ci) are satisfied with S^* in place of T; but then $S^* \in \mathcal{F}'$, and (4.5) contradicts (4.2). Therefore,

(4.6) whenever $S^* \in \mathcal{F}$, $S^* \supseteq P$, $|S^*| \geq |N_S(T)|$, and x and y are elements of S^* (or of $\mathcal{S}(S^*)$) conjugate in $N(S^*)$, then $x \sim y$.

By Lemma 4.1 (a), there exist an element g of G and a well-placed subgroup Q^* of S with respect to W for which $T^g = Q^*$. Let $T^* = W(N_S(Q^*))$. By Lemma 4.1, T^* is a well-placed subgroup of S. By another application of Lemma 4.1,

(4.7)
$$Q^*, T^* \in \mathcal{F}.$$

Since $\mathcal{F}$ also contains T, it follows that $N_S(T)$ and $N_S(Q^*)$ are Sylow p-subgroups of $N(T)$ and $N(Q^*)$ respectively. As $T^g = Q^*$, $N_S(T)^g \subseteq N(Q^*)$. Therefore, there exists $h \in N(Q^*)$ for which $N_S(T)^{gh} = N_S(Q^*)$. Then

(4.8)
$$T^{gh} = Q^{*h} = Q^* \quad \text{and} \quad N_S(T)^{gh} = N_S(Q^*).$$

Moreover,

(4.9)
$$(W(N_S(T)))^{gh} = W(N_S(T)^{gh}) = W(N_S(Q^*)) = T^*.$$

Let $U = N_S(W(N_S(T)))$ and let U_0 be a Sylow p-subgroup of $N(W(N_S(T)))$ that contains U. By (4.7), $N_S(T^*)$ is a Sylow p-subgroup of $N(T^*)$. By (4.9) and the previous argument, there exists $k \in N(T^*)$ for which $N_S(T^*) = U_0^{ghk} \supseteq U^{ghk}$. Let $f = ghk$. Then,

(4.10)
$$U^f \subseteq N_S(T^*)$$

and, by (4.9),

(4.11)
$$(W(N_S(T)))^f = (W(N_S(T)))^{ghk} = T^{*k} = T^*.$$

By (4.10) and (4.1), there exist $S_1, \ldots, S_n \in \mathcal{F}$ and $g_1, \ldots, g_n \in G$ such that $g_i \in N(S_i)$ and $P \subseteq S_i$ $(i = 1, \ldots, n)$, $U \subseteq S_1$ and $U^{g_1 \cdots g_i} \subseteq S_{i+1}$ $(i = 1, \ldots, n-1)$, and $g_1 \cdots g_n = f$. By (4.3), there exist $x, y \in U$ (or $x, y \in \mathcal{S}(U)$) and $z \in N(W(N_S(T)))$ for which $x = y^z$ but $x \not\sim y$. Since

$$|S_1| \geq |U| = |N_S(W(N_S(T)))| \geq |N_S(T)|,$$

(4.6) yields that $x \sim x^{g_1}$, $y \sim y^{g_1}$, and hence $x^{g_1} \not\sim y^{g_1}$. By an induction argument,

$$x^{g_1 \cdots g_i} \not\sim y^{g_1 \cdots g_i} \quad \text{for } i = 1, \ldots, n.$$

Taking $i = n$, we have $x^f \not\sim y^f$. By (4.10), $U^f \subseteq N_S(T^*)$. Therefore, $N_S(T^*)$ (or $\mathcal{S}(N_S(T^*))$) contains x^f and y^f. By (4.11),

$$f^{-1}zf = z^f \in (N(W(N_S(T))))^f = N(T^*).$$

Since $(x^f)^{f^{-1}zf} = x^{zf} = y^f$ and $x^f \not\sim y^f$,

(4.12)
$$\sim \text{ does not contain fusion in } N(T^*).$$

As $T^{gh} = Q^*$ and $N_S(T)^{gh} = N_S(Q^*)$, condition (b) and a similar argument yield that

(4.13)
$$\sim \text{ does not contain fusion in } N(Q^*).$$

We claim that (b) and (cii) are satisfied with T^* and Q^* in place of T and Q. By (4.12), we obtain (b). By (4.7) and (4.13), condition (cii) is satisfied if there exists a subgroup R^* of S for which

(4.14) $R^* \supset P$ and either $Q^* = R^*$ or $Q^* = W(R^*)$.

If T satisfies (ci), let $R^* = Q^*$ and then use (4.4) and (4.8) to show that $P = P^{gh} \subset T^{gh} = Q^* = R^*$, from which (4.14) follows. Now assume that T satisfies (cii) for some subgroups Q and R. Let $R^* = (N_S(Q))^{gh}$. Then $R \subseteq N_S(Q)$ and $T = W''(N_S(Q))$ for some $W'' \in \mathcal{W}$. By (4.8),

$$R^* = (N_S(Q))^{gh} \subseteq (N_S(T))^{gh} = N_S(Q^*) \subseteq S$$

and

$$Q^* = T^{gh} = W''(N_S(Q)^{gh}) = W''(R^*).$$

Moreover, since $R \subseteq N_S(Q)$ and $R \supset P$,

$$R^* = (N_S(Q))^{gh} \supseteq R^{gh} \supset P^{gh} = P.$$

Thus we obtain (4.14) in this case as well. Consequently, in all cases, (b) and (cii) are satisfied with T^* and Q^* in place of T and Q. In particular, $T^* \in \mathcal{F}'$.

By (4.8) and (4.2),

$$|N_S(Q^*)| = |N_S(T)| \geq |N_S(T^*)| = |N_S(W(N_S(Q^*)))|$$

$$\geq |N_S(N_S(Q^*))|.$$

As S is a p-group, $S = N_S(Q^*)$ and $S = N_S(T)$, contrary to (4.4). This contradiction completes the proof of Theorem 4.3.

In order to apply Theorem 4.3, we extend some concepts introduced in §5 of GL.

Definitions. Suppose that $\mathcal{W}$ is a set of conjugacy functors for p on G. Let $H \subseteq G$.

(i) We say that $\mathcal{W}$ *controls strong fusion* in H provided that there exists a Sylow p-subgroup T of H with the following property:

Let $\sim$ be the smallest equivalence relation on $\mathcal{S}(T)$ for which $x \sim y$ whenever x and y are conjugate in $N_H(W(T))$ for some $W \in \mathcal{W}$. Then $x \sim y$ whenever x and y are conjugate in H.

(ii) We say that $\mathcal{W}$ *controls transfer* in H if

$$T \cap H' = \langle T \cap (N_H(W(T)))' \, | \, W \in \mathcal{W} \rangle$$

for some Sylow p-subgroup T of H.

Note that the above definitions do not depend on the choice of T. If $\mathcal{W}$ contains only a single conjugacy functor W, then the above definitions reduce to the definition of control of strong fusion and of transfer by W given in GL. The property in (i) may be stated as follows:

Let $\sim$ be the smallest equivalence relation on $\mathcal{S}(T)$ that contains fusion in $N_H(W(T))$ for all $W \in \mathcal{W}$. Then $\sim$ contains fusion in H.

Alperin and Gorenstein have proved (Theorem 5.5 of GL) that a single conjugacy functor controls various aspects of fusion 'globally', that is, in G, if it controls them 'locally'. By an obvious change in the proof of this theorem, we derive the following result from Theorem 4.3.

Theorem 4.4. *Let $\mathbb{W}$ be a set of conjugacy functors on G. Let $\mathcal{F}$ be the set of all nonidentity subgroups T of S for which $N_S(T)$ is a Sylow p-subgroup of $N(T)$.*

(i) *If $\mathbb{W}$ controls strong fusion in $N(T)$ for every $T \in \mathcal{F}$, then $\mathbb{W}$ controls strong fusion in G.*

(ii) *If $\mathbb{W}$ controls transfer in $N(T)$ for every $T \in \mathcal{F}$, then $\mathbb{W}$ controls transfer in G.*

The following concept was introduced by H. Wielandt.

Definition. Let A be a subgroup of S. Then A is *strongly closed in S (with respect to G)* if

$$A^g \cap S \subseteq A \quad \text{for every } g \in G.$$

Lemma 4.5. *Let $H, K \subseteq G$. Assume that HS is a subgroup of G, $S \subseteq K$, and $G = HK$.*

(a) *Suppose that $T \subseteq S$, $g \in G$, and $T^g \subseteq S$. Then there exist $h \in H$ and $k \in K$ such that $hk = g$ and $T^h \subseteq S$.*

(b) *Suppose that $\sim$ is an equivalence relation on S (or on $\mathcal{S}(S)$) and $\sim$ does not contain fusion in G. Then $\sim$ does not contain fusion in HS or $\sim$ does not contain fusion in K.*

Proof. This is Lemma 6.3 of GL.

Definition. Suppose $H \subseteq G$ and U is a p-subgroup of H. We say that U *controls strong fusion* in H if there exists a Sylow p-subgroup T of H containing U with the following property: Whenever two elements of $\mathcal{S}(T)$ are conjugate in H, they are conjugate in $N_H(U)$.

We note that if the above definition is satisfied for some subgroups H, U, and T, then $U \lhd N_H(T)$ and the definition is satisfied when T is replaced by any Sylow p-subgroup of $N_H(U)$. Moreover, if $\mathbb{W}$ is a conjugacy functor for p on G and $H \subseteq G$, then $\mathbb{W}$ controls strong fusion in H if and only if $\mathbb{W}(T)$ controls strong fusion in H for every Sylow p-subgroup T of H.

Theorem 4.6. *Suppose A is a strongly closed Abelian subgroup of S with respect to G. Then A controls strong fusion in G.*

Proof. We use induction on $|G|$. The result is trivial if $A = 1$, and so we will assume that $A \neq 1$. Since $A \lhd S$, $A \cap Z(S) \supsetneq 1$.

Suppose $Q, R \subseteq S$, $g \in G$, and $Q^g = R$. Then

$$(A \cap Q)^g = A^g \cap R = A^g \cap S \cap R \subseteq A \cap R.$$

Similarly, $(A \cap R)^{g^{-1}} \subseteq A \cap Q$. Therefore,

(4.15) if $Q, R \subseteq S$, $g \in G$, and $Q^g = R$, then $A \cap R = (A \cap Q)^g$.

Define a function $\mathbb{W}$ on the set $\mathcal{C}_p(G)$ of all p-subgroups of G as follows:

$$\mathbb{W}(P^g) = (A \cap P)^g \quad \text{if } P \subseteq S, \, g \in G, \text{ and } A \cap P \neq 1;$$

$$\mathbb{W}(P^g) = P^g \qquad\quad \text{if } P \subseteq S, \, g \in G, \text{ and } A \cap P = 1.$$

Then $\mathbb{W}$ is well defined by Sylow's Theorem, (4.15), and a short argument; in particular, if Q, $R \subseteq S$, g, $h \in G$, and $Q^g = R^h$, then, by (4.15),

$$A \cap R = (A \cap Q)^{gh-1} \quad \text{and} \quad (A \cap R)^h = (A \cap Q)^g.$$

Clearly, $\mathbb{W}$ is a conjugacy functor on G and $\mathbb{W}(S) = A$.

Now define an equivalence relation on $\mathcal{S}(S)$ by

(4.16) $x \sim y$ if and only if x is conjugate to y in $N(A)$.

To prove the conclusion of the theorem, it will be sufficient (and necessary) to show that $\sim$ contains fusion in G. Assume it does not; we will obtain a contradiction. Obviously,

(4.17) $N(A) \subset G$.

Let $\mathbb{\mathcal{W}} = \{\mathbb{W}\}$. Clearly, $\sim$ contains fusion in $N(\mathbb{W}(S))$. Therefore, the hypothesis of Theorem 4.3 is satisfied. Take T as in Theorem 4.3. Then $N_S(T)$ is a Sylow p-subgroup of $N(T)$ and $A \cap N_S(T) \supseteq A \cap Z(S) \supset 1$. By the definition of $\mathbb{W}$,

$$\mathbb{W}(N_S(T)) = A \cap N_S(T) = N_A(T).$$

Since A is strongly closed in S with respect to G, $N_A(T)$ is strongly closed in $N_S(T)$ with respect to $N(T)$. By condition (b) of Theorem 4.3, there exist x, $y \in \mathcal{S}(N_S(T))$ such that x and y are conjugate in $N(T)$ and $x \not\sim y$. By condition (a) of Theorem 4.3, x and y are not conjugate in $N_{N(T)}(N_A(T))$. Therefore, by induction, $N(T) = G$. Thus $T \lhd G$.

Suppose T satisfies condition (cii) of Theorem 4.3. Take Q and R as in (cii). Then $O_p(G) \subset R$, $T = \mathbb{W}(N_S(Q))$, and $Q = R$ or $Q = \mathbb{W}(R)$. Therefore,

(4.18) $N_A(R) \subseteq N_A(Q) = \mathbb{W}(N_S(Q)) = T \subseteq O_p(G) \subset R$.

As $A \lhd T$, AR is a group. Hence, $N_{AR}(R) = N_A(R)R = R$. Since AR is a p-group, $R = AR$. So, $A \subseteq R$. By (4.18), $A = N_A(R) \subseteq T$. Since $T \lhd G$ and $T \subseteq S$, the strong closure of A yields that $A \lhd G$, contrary to (4.17).

Now T must satisfy condition (ci) of Theorem 4.3. Take x, $y \in \mathcal{S}(T)$ for which

(4.19) x is conjugate to y in $N(T)$ and $x \not\sim y$.

By the strong closure of A, we have $A \cap T \lhd G$ and $[T, A] \subseteq T \cap A$. Let $C = C(A \cap T) \cap C(A/(A \cap T))$. Recall that A is Abelian. Then $C \lhd T$ and $A \subseteq C$. By the Frattini argument and the strong closure of A, $G = CN(S \cap C)$, $A \lhd N(S \cap C)$, and $G = CN(A)$.

Obviously, $C(T) \subseteq C$. By Lemma 2.8 and Lemma 2.2 (d),

$$C/C(T) \subseteq O_p(G/C(T)) \subseteq SC(T)/C(T).$$

Consequently, $G = CN(A) = C(T)SN(A) = C(T)N(A)$.

Take x, $y \in \mathcal{S}(T)$ as in (4.19). Then $x \not\sim y$ and $y = x^{hk} = x^k$ for some $h \in C(T)$

and $k \in N(A)$. This contradicts (4.16) and thus completes the proof of Theorem 4.6.

Theorem 4.6 has the following consequence: Assume that $T \lhd S$ and that any two elements of S that are conjugate in G are already conjugate in $N(T)$. Then $Z(T)$ controls strong fusion in G.

5. **Section conjugacy functors.** In §4, we showed that control of fusion or transfer could be proved for the entire group G if it could be proved for all p-local subgroups of G. In this section, we obtain a further reduction, from p-local subgroups to sections G^* with the property that

$$\mathbf{C}_{G^*}(\mathbf{O}_p(G^*)) \subseteq \mathbf{O}_p(G^*).$$

We then apply this reduction to groups in which all such sections G^* possess one factorization or three related factorizations.

As is usual, we will identify H/L with H when $L = 1$ and we will identify the factor group $(H/N)/(K/N)$ with H/K whenever K, $N \lhd H$ and $N \subseteq K$. We will need the concept of a *section conjugacy functor* introduced in §6 of GL.

Definition. Let $\mathcal{C}_p^*(G)$ be the set of all sections of G that are p-groups. A *section conjugacy functor* (for p) is a mapping $\mathbf{W}$ of $\mathcal{C}_p^*(G)$ into $\mathcal{C}_p^*(G)$ that satisfies the following four conditions for every section H/K in $\mathcal{C}_p^*(G)$:

(i) The group $\mathbf{W}(H/K)$ is a subgroup of H/K.

(ii) If $H/K \neq 1$, then $\mathbf{W}(H/K) \neq 1$.

(iii) Let $\mathbf{W}(H/K) = L/K$. Then for all $g \in G$, $\mathbf{W}(H^g/K^g) = L^g/K^g$.

(iv) Suppose that $N \lhd H$ and $N \subseteq K$ and K/N is a p'-group. Let P/N be a Sylow p-subgroup of H/N, and let $\mathbf{W}(P/N) = L/N$. Then $\mathbf{W}(H/K) = LK/K$.

Note that in condition (iv), $H = KP$, $H/K \cong P/N$, and $\mathbf{W}(H/K) \cong \mathbf{W}(P/N)$. Moreover, (iii) and (iv) are satisfied if $\phi(\mathbf{W}(Q)) = \mathbf{W}(R)$ whenever $Q, R \in \mathcal{C}_p^*(G)$ and ϕ is an isomorphism of Q onto R. Thus, all the familiar nonidentity characteristic subgroups of an arbitrary nonidentity p-group P, such as $Z(P)$, $\Omega_1 Z(P)$, and $J(P)$, determine section conjugacy functors.

The next two results, both elementary, are Lemmas 6.1 and 6.2 of GL. The second is proved in page 124 of [39].

Lemma 5.1. *Suppose $\mathbf{W}$ is a section conjugacy functor on G and H/K is a section of G. Then the restriction of $\mathbf{W}$ to $\mathcal{C}_p^*(H/K)$ is a section conjugacy functor on H/K.*

Lemma 5.2 (Dedekind). *Let H, K, $L \subseteq G$. Assume that HL is a subgroup of G and $L \subseteq K$. Then $HL \cap K = (H \cap K)L$.*

Lemma 5.3. *Suppose L is a normal subgroup of G and K is a subgroup of G that contains S. Let $x, y \in S$ or $x, y \in \mathcal{S}(S)$.*

(a) *If L is a p'-group and x and y are conjugate in LK, then x and y are conjugate in K.*

(b) *We have $(SL/L) \cap (KL/L)' = (S \cap K')L/L$.*

(c) *Suppose L is a p-group and T is a subgroup of $S \cap G'$ that contains $[L, G]$ and S'. Assume that $S \cap G' \subseteq LT$. Then $T = S \cap G'$.*

Proof. (a) Take some $g \in G$ for which $x^g = y$. Let $T = \langle x \rangle$ if $x \in S$, and let $T = \langle x_1, \ldots, x_n \rangle$ if $x = (x_1, \ldots, x_n) \in \mathcal{S}(S)$. By Lemma 4.5 (a), there exist $h \in L$ and $k \in K$ for which $hk = g$ and either $x^h \in S$ or $x^h \in \mathcal{S}(S)$.

Suppose $x \in S$. Then $x^h \in S$ and $x^h \equiv x$ (modulo L). Thus $x^h x^{-1} \in S \cap L = 1$ and $x^h = x$. A similar argument applies if $s \in \mathcal{S}(S)$. Hence, in both cases, $y = x^{hk} = x^k$, so that x and y are conjugate in K, as desired.

(b) Obviously

$$(S \cap K')L \subseteq SL \cap K'L \subseteq SL \cap (KL)'L.$$

After factoring by L, we obtain

$$(S \cap K')L/L \subseteq (SL/L) \cap ((KL)'L/L) = (SL/L) \cap (K/L)'.$$

Therefore, it will suffice to prove that $|(SL/L) \cap (KL/L)'| = |(S \cap K')L/L|$.

By Lemma 2.2, SL/L is a Sylow p-subgroup of KL/L and $(SL/L) \cap (KL/L)'$ is a Sylow p-subgroup of $(KL/L)'$. Hence,

$$|(SL/L) \cap (KL/L)'| = |(KL/L)'|_p = |K'L/L|_p = |K'/(K' \cap L)|_p = |K'|_p/|K' \cap L|_p.$$

Similarly,

$$|K'|_p/|K' \cap L|_p = |S \cap K'|/|S \cap K' \cap L| = |(S \cap K')L/L|.$$

Now we have the desired equality.

(c) Since L is a normal p-subgroup of G, we have $L \subseteq S$. By taking factor groups of G, S, L, and T, modulo $[L, G]$, we may assume that $L \subseteq Z(G)$. By Dedekind's Lemma and the hypothesis,

$$T \subseteq S \cap G' = LT \cap (S \cap G') = (L \cap (S \cap G'))T.$$

By Proposition 3.5, $L \cap S \cap G' \subseteq Z(G) \cap S \cap G' \subseteq S' \subseteq T$. Therefore, $T \subseteq S \cap G' \subseteq T$. Hence $T = S \cap G'$.

Lemma 5.4. *Suppose $P = O_p(G)$, $g \in G$, and $R \subseteq S$. Assume that $G = C(P)S$, $g \in N(R)$, g centralizes RP/P, and $N_S(R)$ is a Sylow p-subgroup of $N(R)$. Then there exists $h \in S$ such that gh^{-1} centralizes R.*

Proof. Let $H = N(R)$ and $C = C_H(RP/P)$. Then $C \lhd H$, whence $C \cap S$ is a Sylow p-subgroup of C. Moreover,

$$C_C(P), \; C_C(RP) \lhd H.$$

Since $C_C(P)$ stabilizes the chain $RP \supseteq P \supseteq 1$, Lemma 2.8 yields that $C_C(P)/C_C(RP)$ is a p-group. By hypothesis, $G/C(P)$ is a p-group. Therefore, $C/C_C(P)$ and $C/C_C(RP)$ are p-groups, and

$$C = (C \cap S)C_C(RP).$$

Since $g \in C$, we are done.

Now we come to our main reduction.

Theorem 5.5. *Let $\mathbb{W}$ be a set of section conjugacy functors on G.*

(a) *Assume that whenever G^* is a section of G, S^* is a Sylow p-subgroup of G^*, and $C_{G^*}(O_p(G^*)) \subseteq O_p(G^*)$, then $\mathbb{W}$ controls strong fusion in G^*. Then $\mathbb{W}$ also controls strong fusion in G.*

(b) *Assume that whenever G^* is a section of G, S^* is a Sylow p-subgroup of G^*, and $C_{G^*}(O_p(G^*)) \subseteq O_p(G^*)$, then $\mathbb{W}$ controls transfer in G^*. Then $\mathbb{W}$ also controls transfer in G.*

Proof. The proofs of (a) and (b) are close enough that we may give them simultaneously, except at a few points. Where the proofs differ, we shall generally state the argument that applies to (a) and then state the appropriate substitution for (b) in parentheses. To a large extent, we follow the proofs of the analogous results in GL, namely, Theorems 6.6 and 6.8.

We use induction on $|G|$. By Lemma 5.1, $\mathbb{W}$ controls strong fusion (transfer) in every section of G except possibly G itself. By Theorem 4.4, we may assume that $O_p(G) \neq 1$.

Let $\sim$ be the smallest equivalence relation on $\delta(S)$ that contains fusion in $N(W(S))$ for every $W \in \mathbb{W}$. (Let

$$S^* = \langle S \cap (N(W(S)))' \,|\, W \in \mathbb{W} \rangle$$

and let $\sim$ be the equivalence relation on S given by $x \sim y$ if $xy^{-1} \in S^*$.) Then

(5.1) (for (a)) if $x, y \in \delta(S)$ and $x \sim y$, then x and y are conjugate in G;

(5.2) (for (b)) if $x, y \in S$ and $x \sim y$, then $xy^{-1} \in S^* \subseteq S \cap G'$.

By the definition of control of strong fusion (by the Focal Subgroup Theorem and the definition of control of transfer),

(5.3) $\mathbb{W}$ controls strong fusion (transfer) in G if $\sim$ contains fusion in G.

Similarly,

(5.4) whenever $S \subseteq H \subset G$, then $\sim$ contains fusion in H.

Let $P = O_p(G)$, $C = C(P)$, and $R = PC \cap S$. Since $PC \lhd G$, the Frattini argument yields that

$$G = PCN(R) = CPN(R) = CN(R).$$

By (5.3), (5.4) and Lemma 4.5, we may assume that

(5.5) $G = CS$ or $G = N(R)$.

We will first treat the case in which $G = N(R)$. Therefore, *until further notice, assume that* $G = N(R)$. Then $R \subseteq O_p(G) = P$ and $C_S(P) \subseteq R \subseteq P$. Let $L = O_{p'}(G)$. By Lemma 3.8,

(5.6) $C_{G/L}(O_p(G/L)) \subseteq O_p(G/L)$.

By part (iv) of the definition of a section conjugacy functor (with 1, S, L, LS, and $\mathbb{W}(S)$ in place of N, P, K, H, and L),

(5.7) $\mathbb{W}(SL/L) = \mathbb{W}(S)L/L$ for every $\mathbb{W} \in \mathbb{\hat{W}}$.

By an application of the Frattini argument,

(5.8) $N_{G/L}(\mathbb{W}(SL/L)) = N(\mathbb{W}(S))L/L$ for every $\mathbb{W} \in \mathbb{\hat{W}}$.

Now let us use bars to denote images of elements, groups, and sequences of elements when G is projected onto G/L. Define an equivalence relation $\approx$ on $\mathcal{S}(\overline{S})$ (on $\overline{S}$) as follows:

$$\text{for } x, y \in \mathcal{S}(S) \ (x, y \in S), \quad \overline{x} \approx \overline{y} \text{ if } x \sim y.$$

Since

(5.9) each coset of L in LS contains a unique element of S, $\approx$ is well-defined. Note that, by Sylow's Theorem,

(5.10) whenever $z \in S$, then z is conjugate in $\langle z \rangle L$ to every p-element of the coset zL.

Now suppose $x, y \in \mathcal{S}(S)$ (or $x, y \in S$), $\mathbb{W} \in \mathbb{\hat{W}}$, and $\overline{x}$ and $\overline{y}$ are conjugate in $N(\mathbb{W}(\overline{S}))$. By (5.8), (5.9), and (5.10), x and y are conjugate in $LN(\mathbb{W}(S))$. By Lemma 5.3, x and y are conjugate in $N(\mathbb{W}(S))$. By the definition of $\sim$, we have $x \sim y$ and hence $\overline{x} \approx \overline{y}$. Thus,

(5.11) whenever $x, y \in \mathcal{S}(S)$ (or $x, y \in S$), $\mathbb{W} \in \mathbb{\hat{W}}$, and $\overline{x}$ and $\overline{y}$ are conjugate in $N_{\overline{G}}(\mathbb{W}(\overline{S}))$, then $\overline{x} \approx \overline{y}$.

Now assume $u, v \in \mathcal{S}(S)$ (or $u, v \in S$), and u and v are conjugate in G. Then $\overline{u}$ and $\overline{v}$ are conjugate in $\overline{G}$. For the proof of (a), (5.11) and the hypothesis of the theorem yield that $\overline{u} \approx \overline{v}$. Then $u \sim v$ by the definition of $\approx$. For the proof of (b), we take S^* as defined just before (5.1) and use (5.8) and Lemma 5.3 to show that

$$\langle \overline{S} \cap (N(\mathbb{W}(\overline{S})))' | \mathbb{W} \in \mathbb{\hat{W}} \rangle \subseteq \overline{S}^*;$$

by the hypothesis of the theorem, $\overline{uv^{-1}} \in \overline{S}^*$, whence $uv^{-1} \in S^*L \cap S = S^*$ and $u \sim v$. Thus, for both proofs, $u \sim v$. Since u and v are arbitrary, (5.3) yields the desired result.

Now, *we assume that* $G \neq N(R)$. By (5.5), $G = CS$. Henceforth, we will use bars to denote images modulo P, rather than images modulo L. We will also define the relation $\approx$ differently. For $u, v \in \mathcal{S}(\overline{S})$ (or $u, v \in \overline{S}$), we say that $u \approx v$ if the following condition is satisfied:

Whenever $x, y \in \mathcal{S}(S)$ (or $x, y \in S$), $\overline{x} = u$, and $\overline{y} = v$, then there exist $x', y' \in \mathcal{S}(S)$ (or $x', y' \in S$) for which $\overline{x'} = u$, $\overline{y'} = v$, $x' \sim y$, and $x \sim y'$.

It is not difficult to see that $\approx$ is an equivalence relation on $\mathcal{S}(S)$ (or on S).

Take any $\mathbb{W} \in \mathbb{\hat{W}}$. Since $1 \subset O_p(G) = P \subseteq PC \cap S = R$ and $G \neq N(R)$,

$$1 \subset O_p(G) = P \subset S.$$

Hence $\bar{S} \neq 1$, $W(\bar{S}) \neq 1$, and $N_{\underset{G}{}}(W(\bar{S})) \subset \bar{G}$. Suppose $h \in N_{\underset{G}{}}(W(\bar{S}))$ and u, $u^h \in \mathcal{S}(\bar{S})$ (or u, $u^h \in \bar{S}$). Take g, x, and y such that $g \in G$, x, $y \in \mathcal{S}(S)$ (or x, $y \in S$), $\bar{g} = h$, $\bar{x} = u$, and $\bar{y} = u^h$. Then x^g, $y^{g^{-1}} \in \mathcal{S}(S)$ (or x^g, $y^{g^{-1}} \in S$), $\overline{x^g} = u^h$, and $\overline{y^{g^{-1}}} = u$. By (5.4) applied to the inverse image of $N_{\underset{G}{}}(W(\bar{S}))$ in G, we have $y^{g^{-1}} \sim y$ and $x \sim x^g$. Consequently, $u \approx u^h$. This shows that

(5.12) $u \approx u^h$ whenever $W \in \mathfrak{W}$, $h \in N_{\underset{G}{}}(W(\bar{S}))$, and u, $u^h \in \mathcal{S}(\bar{S})$ (or u, $u^h \in \bar{S}$).

Now we separate the proofs of (a) and (b). For (a), suppose $g \in G$, x, $y \in \mathcal{S}(S)$, and $x^g = y$. Then $\bar{x}$ and $\bar{y}$ are conjugate in $\bar{G}$. Since $P \neq 1$, $|\bar{G}| < |G|$. By (5.12) and induction,

(5.13) $$\bar{x} \approx \bar{y}.$$

Let $y = (y_1, \ldots, y_n)$ and let $U = \langle y_1, \ldots, y_n, P \rangle$. Take $h \in G$ such that
(5.14) S contains a Sylow p-subgroup of $(N(U))^h$.
Let $z = y^h$ and $R = U^h$. Then $z = x^{gh}$. By the proof of (5.13), $\bar{x} \approx \bar{z}$. By the definition of $\approx$, there exists some $z' \in \mathcal{S}(S)$ for which $z' \sim x$ and $\bar{z'} = \bar{z}$. By (5.1), $x^k = z'$ for some $k \in G$. Then

$$z = x^{gh} = z'^{k^{-1}gh}.$$

As $\bar{z'} = \bar{z}$, $k^{-1}gh$ centralizes RP/P. By (5.14) and Lemma 5.4, there exists $t \in S$ such that

$$z'^t = z'^{k^{-1}gh} = z.$$

Then $z' \sim z$ by taking $G^* = S$ in the hypothesis. Since $x \sim z'$, $x \sim z$. Similarly, $y \sim z$. It follows immediately that $x \sim y$. Since x and y are arbitrary elements of $\mathcal{S}(S)$ conjugate in G, this shows that $\sim$ contains fusion in G in the proof of (a). By (5.3), this completes the proof of (a).

Now we complete the proof of (b). Recall that here, by definition,

(5.15) $$S^* = \langle S \cap (N(W(S)))' \mid W \in \mathfrak{W} \rangle.$$

By (5.12) and the definitions of $\sim$ and $\approx$, $u^h u^{-1} \in \overline{S^*}$ whenever $W \in \mathfrak{W}$, $h \in N_{\underset{G}{}}(W(\bar{S}))$ and u, $u^h \in \bar{S}$. Since $|\bar{G}| < |G|$, the Focal Subgroup Theorem and induction now yield that $\bar{S} \cap (\bar{G})' \subseteq \overline{S^*}$. By (5.2), $S^* \subseteq S \cap G'$. Hence,

$$S^* \subseteq S \cap G' \subseteq S^*P.$$

Since $G = CS$,

$$[P, G] = [P, S] \subseteq S'.$$

By putting $G^* = S$ in the hypothesis, we see that $[P, G] \subseteq S' \subseteq S^*$. By Lemma 5.3(c) (with G, P, and S^* in place of K, L, and T), $S^* = S \cap G'$. This completes the proof of (b) and of Theorem 5.5.

Corollary 5.6. *Suppose* W_1 *and* W_2 *are section conjugacy functors on* G. *Assume that whenever* G^* *is a section of* G, S^* *is a Sylow* p-*subgroup of* G^*, *and* $C_{G^*}(O_p(G^*)) \subseteq O_p(G^*)$, *then*

$$G^* = N_{G^*}(W_1(S^*))N_{G^*}(W_2(S^*)).$$

Then $\{W_1, W_2\}$ *controls strong fusion in* G.

Proof. By Theorem 5.5, it suffices to show that $\{W_1, W_2\}$ controls strong fusion in G^* for every section G^* as above. However, this follows easily from Lemma 4.5(a) and the definition of control of strong fusion.

Corollary 5.7. *Suppose* W_1 *and* W_2 *are section conjugacy functors on* G. *Assume that whenever* G^* *is a section of* G, S^* *is a Sylow* p-*subgroup of* G^*, *and* $C_{G^*}(O_p(G^*)) \subseteq O_p(G^*)$, *then*

$$G^* = C_{G^*}(\Omega_1 Z(S^*))N_{G^*}(W_1(S^*)) = C_{G^*}(\Omega_1 Z(S^*))N_{G^*}(W_2(S^*))$$

$$= N_{G^*}(W_1(S^*))N_{G^*}(W_2(S^*)).$$

Assume also that $\Omega_1 Z(S) \subseteq W_1(S) \cap W_2(S)$.
  Let

$$A = \langle \Omega_1 Z(S)^x | x \in N(W_1(S)) \rangle.$$

Then A *is an elementary Abelian strongly closed subgroup of* S *with respect to* G *and* A *controls strong fusion in* G.

Proof. Let $W_1 = W_1(S)$, $W_2 = W_2(S)$, and $V = \Omega_1 Z(W_1)$. The hypothesis yields that

(5.16) $A \lhd N(W_1)$

and $\Omega_1 Z(S) \subseteq V$. Therefore, $A \subseteq V$, and A is elementary Abelian.
  Let $C = C(V) \cap N(W_1)$, $H = N(W_1) \cap N(C \cap S)$, and $G^* = H/O_{p'}(H)$. By the Frattini argument,

(5.17) $N(W_1) = CH$.

Moreover, $S \subseteq H$ and $C_S(O_p(H)) \subseteq C_S(V) \subseteq C \cap S \subseteq O_p(H)$. By Lemma 3.8, $C_{G^*}(O_p(G^*)) \subseteq O_p(G^*)$. By the hypothesis, the definition of a section conjugacy functor, and the Frattini argument,

(5.18) $H = O_{p'}(H)C_H(\Omega_1 Z(S))N_H(W_2)$.

  Let $B = \langle \Omega_1 Z(S)^x | x \in N(W_2) \rangle$ and $K = C(\Omega_1 Z(S)) \cap N(W_1)$. Since $O_{p'}(H)$ centralizes $O_p(H)$, it centralizes V. Hence, $O_{p'}(H) \subseteq C$. In addition, $C \subseteq K$. Therefore, by (5.17) and (5.18),

$$N(W_1) = KN(W_1) = KCO_{p'}(H)C_H(\Omega_1 Z(S))N_H(W_2) = KN_H(W_2)$$

and

$$A = \langle \Omega_1 Z(S)^x | x \in N(W_1) \rangle = \langle \Omega_1 Z(S)^y | y \in N_H(W_2) \rangle \subseteq B.$$

By symmetry, $B \subseteq A$. Thus, $A = B$.

By (5.16) and symmetry, A is normalized by both $N(W_1)$ and $N(W_2)$. By Corollary 5.6 and the hypothesis, $\{W_1, W_2\}$ controls strong fusion in G. Therefore, A is strongly closed in S with respect to G. By Theorem 4.6, A controls strong fusion in G.

§6 and the remainder of §5 are devoted to transfer. We shall require some notation from the appendix of GL for the situation when G acts as an operator group on a group V. (This includes the situation in which $V \lhd G$ and G acts on V by conjugation.) For $g \in G$, $v \in V$, and all $i \geq 1$,

$$[v, g; 1] = [v, g] = v^{-1} v^g \quad \text{and} \quad [v, g; i+1] = [[v, g; i], g].$$

For $W \subseteq V$, $g \in G$, and $H \subseteq G$, $[W, g] = \langle [w, g] | w \in W \rangle$ and $[W, H] = \langle [W, h] | h \in H \rangle$. Finally, for $g \in G$ and all $i \geq 0$, $[V, g; 0] = V$ and $[V, g; i+1] = [[V, g; i], g]$.

As usual, for a subgroup H of G, a *transversal* of H in G is a subset T of G for which G is the disjoint union of the sets Ht, $t \in T$.

The next result is part of Lemmas A1.7 and A1.8 of GL.

Lemma 5.8. *Suppose V is an elementary Abelian p-group and G is an operator group on V. Let $g \in S$. Then*

(a) $\Pi_{0 \leq i \leq p-1} v^{g^i} = [v, g; p-1]$, *for all $v \in V$; and*

(b) *if $[V, g; p-1] = 1$, then for every proper subgroup A_1 of $\langle g \rangle$ and every transversal R to A_1 in $\langle g \rangle$,*

$$\prod_{h \in R} v^{h-1} = 1 \quad \text{for all } v \in V.$$

The following proposition is a variant of some results of W. Feit, L. Scott, and the author (GL, p. 21). Recall that an element or subgroup W of S is *weakly closed* in S (with respect to G) if $W^x = W$ whenever $x \in G$ and W^x is contained in S.

Proposition 5.9. *Suppose V is a finite p-group and G is an operator group on V. Let $\mathfrak{A}$ be a nonempty set of subgroups of S and let B be the subgroup of S generated by the elements of $\mathfrak{A}$. Suppose M is a subgroup of G and $N(B) \subseteq M$. Assume that $\mathfrak{A}$ satisfies the following conditions:*

(a) *B is weakly closed in S with respect to G, and*

(b) *whenever $A \in \mathfrak{A}$, $g \in G - M$, $A \cap S^g = A \cap M^g \subset A$, W is a composition factor of V under G, and $v \in W$, then*

$$\prod_{h \in R} v^{h-1} \equiv 1 \quad \text{modulo } [W, A \cap S^g],$$

for every transversal R of $A \cap S^g$ in A.

Then $B \lhd S$ and $[V, G] = [V, M]$.

Proof. This result extends slightly Theorem A1.6 of GL and requires only small changes in its proof (and in the statement and proof of the preceding result, Theorem A1.4). We note that these proofs may be simplified by using the methods and results of §6 (in particular, Proposition 6.2 and Proposition 6.6 with $R = W$).

Theorem 5.10. *Suppose $\mathbb{W}$ is a nonempty set of section conjugacy functors on G. Assume that $\mathbb{W}$ controls transfer in every section of G other than G itself, but that $\mathbb{W}$ does not control transfer in G. Let*

$$T = \langle S \cap (N(W(S)))' | W \in \mathbb{W} \rangle \quad and \quad M = N(T \cap 0_p(G)).$$

Then the following conditions are satisfied:

(a) $C(0_p(G)) \subseteq 0_p(G)$;

(b) M *is the unique maximal subgroup of G that contains S;*

(c) $[0_p(G), M] \subseteq 0_p(G) \cap M' \subseteq 0_p(G) \cap T$; *and*

(d) *for every $u \in S - 0_p(G)$, there exists a chief factor X/Y of G for which* $[X/Y, u; p - 1] \neq 1$.

Moreover, suppose A_0 is a subgroup of S that is not contained in $0_p(G)$. Then there exist $A \subseteq S$, $g \in G - M$, a transversal R of $A \cap S^g$ in S, a chief factor X/Y of G, and an element v of X/Y for which A is conjugate to A_0 in G, $A \cap S^g = A \cap M^g \subset A$, $X \subseteq 0_p(G)$, and

$$\prod_{h \in R} v^{h-1} \neq 1 \quad modulo \ [X/Y, A \cap S^g].$$

Proof. By Theorem 5.5, G satisfies (a).

Take any $W \in \mathbb{W}$. Then

$$(5.19) \qquad\qquad S' \subseteq S \cap (N(S))' \subseteq S \cap (N(W(S)))' \subset T.$$

By the hypothesis and the definition of control of transfer, $T \neq S \cap G'$. Therefore,

$$(5.20) \qquad\qquad S' \subseteq T \subset S \cap G' \quad and \quad N(S) \subset G.$$

By Lemma 5.3 (c) (with $L = P$),

$$(5.21) \qquad\qquad [P, G] \not\subseteq T \quad or \quad S \cap G' \not\subseteq PT.$$

We will show that $S \cap G' \subseteq PT$. Let $\bar{G} = G/P$ and, for every subgroup H of G, let $\bar{H} = HP/P$. Take any $W_0 \in \mathbb{W}$ and take the subgroup K of G for which $K \supseteq P$ and $K/P = N_{\bar{G}}(W_0(\bar{S}))$. By (5.20), $N(S) \subset G$. Therefore, $\bar{S} \neq 1$ and $W_0(\bar{S}) \neq 1$. Since $W_0(\bar{S}) \lhd \bar{S}$ and $0_p(\bar{G}) = 1$, we have $S \subseteq K \subset G$. By hypothesis,

$$S \cap K' = \langle S \cap (N_K(W(S)))' | W \in \mathbb{W} \rangle \subseteq T.$$

Therefore, by Lemma 5.3 (b) (with $L = P$),

$$\bar{S} \cap (N_{\bar{G}}(W_0(\bar{S})))' = \bar{S} \cap \bar{K}' = (S \cap K')P/P \subseteq TP/P.$$

Since W_0 was chosen arbitrarily in $\mathfrak{W}$ and $|\bar{G}| < |G|$,

$$\bar{S} \cap \bar{G}' = \langle \bar{S} \cap (N_{\underset{G}{}}(W(\bar{S})))' \mid W \in \mathfrak{W} \rangle \subseteq TP/P.$$

Consequently, $(S \cap G')P/P \subseteq \bar{S} \cap \bar{G}' \subseteq TP/P$ and $S \cap G' \subseteq TP$, as claimed. By (5.21),

(5.22) $$[P, G] \not\subseteq T.$$

Let $M_0 = C_M(P/(T \cap P))$. By (5.19),

$$P \cap T \supseteq P \cap S \cap (N(S))' = P \cap (N(S))' \supseteq [P, N(S)].$$

Therefore, $N(S) \subseteq M$ and $N(S) \subseteq M_0$. By the Frattini argument, $M = M_0 N(S) = M_0$. Thus, $[P, M] \subseteq T \cap P$. By (5.22),

(5.23) $$M \subset G.$$

Suppose $S \subseteq H \subset G$. By hypothesis,

$$S \cap H' = \langle S \cap (N_H(W(S)))' \mid W \in \mathfrak{W} \rangle \subseteq T.$$

Therefore, $P \cap T \supseteq P \cap S \cap H' = P \cap H' \supseteq [T \cap P, H]$, and H normalizes $T \cap P$. Hence, $H \subseteq M$. Thus, (5.23) yields (b). By taking $H = M$, we obtain (c).

Now we will prove the last part of the conclusion and then derive (d) from it. Assume $A_0 \subseteq S$ and $A_0 \not\subseteq P$. Let $\mathfrak{U}$ be the set of all subgroups of S that are conjugate to A in G and let $B = \langle \mathfrak{U} \rangle$. Then B is a weakly closed subgroup of S with respect to G. Hence, $S \subseteq N(B)$. As $A \subseteq B \subseteq O_p(N(B))$, we have $N(B) \subset G$. By (b), $N(B) \subseteq M$. By (c) and (5.22),

$$[P, N(B)] \subseteq [P, M] \subseteq T \cap P \quad \text{and} \quad [P, G] \not\subseteq T \cap P.$$

Therefore, $[P, M] \neq [P, G]$.

Now, G acts as an operator group on P by conjugation. The above paragraph shows that the commutator condition in Proposition 5.9 must be violated, and this yields the last part of the conclusion. To obtain (d), consider the special case in which $A_0 = \langle u \rangle$. In this case, A is a cyclic subgroup of S, R is a transversal of the proper subgroup $A \cap S^g$ in A, and $\Pi_{h \in R} v^{h-1} \neq 1$. Take $z \in G$ such that $\langle u^z \rangle = A$. By Lemma 5.8, $[X/Y, u^z; p - 1] \neq 1$. This yields (d) and completes the proof of the theorem.

6. Further results on transfer. In a fundamental paper on transfer [64], H. Wielandt introduced the concept of strongly and weakly closed subgroups and proved that under suitable conditions

(6.1) $$S \cap G' = S \cap (N(W))'$$

for some weakly closed subgroup W of S. Recently, T. Yoshida has introduced some important further techniques. Most of the results of Wielandt and Yoshida cannot be obtained by using our general approach of conjugacy functors and reductions. In this section, we give a sample of their methods, including two new cases of (6.1).

We will use the definitions of transversal and extended commutators introduced just before Lemma 5.8. In addition, we will need to refer to an arbitrary Sylow p-subgroup of the symmetric group S^{p^2}. Since any such p-group is known to be isomorphic to the 'wreath product' of Z_p by Z_p [39, pp. 81–83], we will denote it by $Z_p \wr Z_p$.

For an arbitrary subgroup H of G and an arbitrary transversal T to H in G, define a function $\phi: G \to T$ by $H\phi(x) = Hx$, $x \in G$. Recall that the transfer mapping $V(G, H)$ is a homomorphism of G into H/H' given by [39, pp. 200–203]

$$V(G, H)(x) = \prod_{t \in T} tx(\phi(tx))^{-1} \bmod H'.$$

This mapping is independent of the choice of T. When convenient, it can be regarded as a mapping from G/G' into H/H' or, when $H \subseteq K$, as a mapping into K/K'. The former enables us to obtain

$$V(G^*, H)(x) = V(G, H)(V(G^*, G)(x))$$

if $G^* \supseteq G$ and $x \in G^*$.

The main results of this section were originally proved by Yoshida [68] by using character theory and were discussed and applied by K. Harada in a lecture [69] at the Duluth Conference. We will give an unpublished account of this work by I. M. Isaacs, circulated at the conference, which uses only transfer theory.

Lemma 6.1. *Suppose $S \subseteq H \subseteq G$ and X/H' is the image of $V(G, H)$ in H/H'. If $S \cap G' \neq S \cap H'$, then $S \not\subseteq X$.*

Proof. Assume $S \cap G' \neq S \cap H'$. Then $S \cap H' \subseteq S \cap G'$. Since X/H' is Abelian, Ker $V(G, H) \supseteq G'$. Therefore,

$$|X/H'|_p = |\mathrm{Im}\ V(G, H)|_p = |G/\mathrm{Ker}\ V(G, H)|_p \leq |G/G'|_p$$
$$= |S|/|S \cap G'| < |S/(S \cap H')| = |H/H'|_p.$$

Hence, $|X|_p < |H|_p$, and $S \not\subseteq X$.

The following result is analogous to the Mackey Decomposition Theorem [17, pp. 107–108], for representations of groups on modules.

Proposition 6.2. *Suppose H, $K \subseteq G$ and T is a complete set of representatives for the double cosets HxK in G. Let $x \in K$. Then*

$$V(G, H)(x) = \prod_{t \in T} tV(K, K \cap H^t)(x)t^{-1} \bmod H'.$$

Proof. We must first verify that the expression on the right yields a well-defined element of H/H'. Take any $t \in T$. Then $V(K, K \cap H^t)(x)$ belongs to $(K \cap H^t)/(K \cap H^t)'$

and hence determines a unique coset of $(H^t)'$ in H^t. Therefore, $t^{-1}V(K, K \cap H^t)t$ determines a unique coset of H' in H.

For each $t \in T$, let U_t be a transversal of $K \cap H^t$ in K. Let

$$R = \{tu \mid t \in T, u \in U_t\}.$$

Then R is a transversal to H in G because, for each $t \in T$, $\{Htu \mid u \in U_t\}$ is the set of all right cosets of H in HtK. (Consider the right cosets of H^t in H^tK.) Define functions $\phi: G \to R$, $\phi_t: K \to U_t$ $(t \in T)$ as in the definition of the transfer mapping. Then

$$V(G, H)(x) = \prod_{r \in R} rx\phi(rx)^{-1} = \prod_{t \in T} \prod_{u \in U_t} tux\phi(tux)^{-1} \bmod H'$$

$$= \prod_{t \in T} \prod_{u \in U_t} tux(t\phi_t(ux))^{-1} = \prod_{t \in T} \prod_{u \in U_t} tux\phi_t(ux)^{-1}t^{-1} \bmod H'$$

$$= \prod_{t \in T} tV(K, K \cap H^t)(x)t^{-1} \bmod H'.$$

Corollary 6.3. *Suppose $H \subseteq G$ and $x \in G$. Then there exist a subset T of G and natural numbers $r(t)$, $t \in T$, such that*

(a) $tx^{r(t)}t^{-1} \in H$ *for each $t \in T$,*

(b) $\sum_{t \in T} r(t) = |G:H|$, *and*

(c) $V(G, H)(x) = \prod_{t \in T} tx^{r(t)}t^{-1} \bmod H'$.

Proof. Let $K = \langle x \rangle$ and take T as in Proposition 6.2. A short calculation yields that, for each $t \in T$,

$$V(K, K \cap H^t)(x) = x^{r(t)} \quad \text{for } r(t) = |K: K \cap H^t|.$$

This yields (a) and (c). In addition, (b) follows because, for each $t \in T$, $r(t)$ is the number of cosets Hy in HtK.

Lemma 6.4. *Suppose R is a p-group, A is a normal elementary Abelian subgroup of R, $x \in A$, and $z \in R - A$. Assume that*

$$|R/A| = p \quad \text{and} \quad \prod_{0 \le i \le p-1} x^{z^i} \ne 1.$$

Then $Z_p \wr Z_p$ is a homomorphic image of R.

Proof. Let $u = \prod_{0 \le i \le p-1} x^{z^i}$ and let B be a complement to $\langle u \rangle$ in A. Since $|R:B| = |R:A||A:B| = p^2$, the permutation representation of R on the cosets By $(y \in R)$ yields a homomorphism $\phi: R \to S^{p^2}$. As $u \notin B$, Lemma 5.8 yields that

$$[\phi(x), \phi(z); p-1] = \prod_{0 \le i \le p-1} \phi(x)^{\phi(z)^i} = \phi(u) \ne 1.$$

Therefore, Im ϕ has nilpotence class at least p. Consequently,

$$|\text{Im } \phi| \geq p^{p+1} = |S^{p^2}|_p.$$

Since $Z_p \wr Z_p$ is a Sylow p-subgroup of S^{p^2}, $\text{Im } \phi \cong Z_p \wr Z_p$.

Proposition 6.5 (Yoshida). *Suppose R is a p-group, $Q \subset R$, $x \in R$, and $M \triangleleft Q$. Assume that*

(a) *Q/M is elementary Abelian,*

(b) *$V(R, Q)(x) \notin M/Q'$, and*

(c) *no element of $Q - M$ is conjugate in R to a power of x^p.*

Then $R' \cap Q \not\subseteq M$. Moreover, for every maximal subgroup A of R that contains Q, A contains x and, for all $z \in R - A$, $\prod_{0 \leq i \leq p-1} x^{z^i} \notin \Phi(A)$.

Proof. Let $n = |R{:}Q|$. Since R is a p-group, a short argument by induction yields that $x^n \in Q$. By (c) and (a),

(6.2) $\qquad\qquad x^n \in M \subseteq Q$ and $M \supseteq \Phi(Q) \supseteq Q'$.

By Corollary 6.3, there exists $y \in R'$ such that

$$V(R, Q)(x) = x^n y \bmod Q'.$$

As $\text{Im } V(R, Q) \subseteq Q/Q'$, $x^n y \in Q$. Therefore, by (b) and (6.2), $y \in Q - M$. Since $y \in R'$, it follows that $Q \cap R' \not\subseteq M$.

Now suppose A is a maximal subgroup of R that contains Q. Then $A \triangleleft R$ and $|R/A| = p$. Take $u \in A$ such that $uA' = V(R, A)(x)$. Then

$$V(R, Q)(x) = V(A, Q)(V(R, A)(x)) = V(A, Q)(u).$$

Therefore, by (b),

(6.3) $\qquad\qquad V(A, Q)(u) \notin M/Q'.$

Assume first that $x \in R - A$. Then $1, x, \ldots, x^{p-1}$ is a transversal to A in R, and the definition of transfer yields that

$$u = V(R, A)(x) = x^p \bmod A'.$$

By (6.3), $V(A, Q)(x^p) \notin M/Q'$. By Corollary 6.3, some conjugate of a power of x^p lies in $Q - M$. But this contradicts (c). Hence, $x \in A$.

Take $z \in R - A$. Since $x \in A$ and $A \triangleleft R$, the definition of transfer yields that

$$u = V(R, A)(x) = \prod_{0 \leq i \leq p-1} x^{z^i} \bmod A'.$$

Since $V(A, Q)$ maps $\Phi(A)$ into $\Phi(Q)/Q'$, we have $u \notin \Phi(A)$ by (6.2) and (6.3). This completes the proof of Proposition 6.5.

Proposition 6.6. *Suppose W is a weakly closed subgroup of S with respect to G,*

$x \in S$, and H, Y, $R \subseteq G$. Assume that $N(W) \subseteq H \subseteq G$, $H' \subseteq Y \subseteq H$, and $W \subseteq R \subseteq S$. Assume further that $x \in R - Y$ and $V(G, H)(x) \in Y/H'$. Then there exists $t \in G - H$ such that

(a) $R \not\subseteq H^t$, and

(b) $V(R, R \cap H^t)(x) \notin (R \cap Y^t)/(R \cap H^t)'$.

Proof. Let T be a complete set of representatives for the double cosets HyR in G. By Proposition 6.2 and the hypothesis,

$$(6.4) \qquad\qquad H' \subseteq Y \subseteq H, \qquad x \in R - Y,$$

and

$$(6.5) \qquad\qquad 1 = V(G, H)(x) = \prod_{t \in T} tV(R, R \cap H^t)(x)t^{-1} \bmod Y.$$

Let u be the element of T in $H1R$, i.e., in H. By (6.4),

$$uV(R, R \cap H^u)(x)u^{-1} \equiv V(R, R)(x) \equiv x \not\equiv 1 \bmod Y.$$

Therefore, by (6.5), there exists $t \in T - H$ such that

$$tV(R, R \cap H^t)(x)t^{-1} \not\equiv 1 \bmod Y.$$

Then $V(R, R \cap H^t)(x)$ is an element of $(R \cap H^t)/(R \cap H^t)'$ which does not map into $Y^t/(H^t)'$ in the natural projection of $(R \cap H^t)/(R \cap H^t)'$ into $H^t/(H^t)'$. This proves (b).

Suppose $R \subseteq H^t$. Then $W \subseteq R \subseteq H^t$ and $W^{t^{-1}} \subseteq H$. As $S \subseteq H$, there exists $h \in H$ such that $W^{t^{-1}h} \subseteq S$. As W is weakly closed in S, $W^{t^{-1}h} = W$ and $t^{-1}h \in N(W) \subseteq H$. Then $t \in H$, contrary to the choice of t. This proves (a).

Theorem 6.7 (Yoshida). *Suppose W is a weakly closed subgroup of S and $S \cap G' \neq S \cap (N(W))'$. Then there exist $x \in S$, $z \in W$, and a maximal subgroup A of $\langle W, x \rangle$ such that $x \in A$ and*

$$\prod_{0 \le i \le p-1} x^{z^i} \equiv [x, z; p-1] \not\equiv 1 \bmod \Phi(A).$$

Proof. Let $H = N(W)$ and let X/H' be the image of $V(G, H)$ in H/H'. Then $S \cap G' \neq S \cap H'$. By Lemma 6.1, $S \not\subseteq X$. Hence, there exists a (normal) subgroup Y/X of index p in H/X.

Take $x \in S - Y$ of minimal possible order. Let $R = \langle W, x \rangle$. Then $V(G, H)(x) \in X/H' \subseteq Y/H'$. By Proposition 6.6, there exists $t \in G - H$ such that

$$(6.6) \qquad R \not\subseteq H^t \quad \text{and} \quad V(R, R \cap H^t)(x) \notin (R \cap Y^t)/(R \cap H^t)'.$$

Therefore, $R \cap H^t \supset R \cap Y^t$. Since $|H/Y| = p$,

$$(6.7) \qquad\qquad |(R \cap H^t)/(R \cap Y^t)| = p.$$

Now let $Q = R \cap H^t$ and $M = R \cap Y^t$. By (6.6), $Q \subset R$. Let A be a maximal subgroup of R that contains Q. By (6.7), (6.6), and the choice of x, we obtain the hypothesis of Proposition 6.5. Therefore, $x \in A$ and, for all $z \in R - A$,

$$\prod_{0 \le i \le p-1} x^{z^i} \notin \Phi(A).$$

Then $R = \langle W, x \rangle = \langle W, A \rangle$ and $W \nsubseteq A$. By choosing $z \in W - A$ and applying Lemma 5.8 with $V = A/\Phi(A)$, we obtain the conclusion of the theorem.

Now we obtain the new results mentioned at the beginning of the section.

Theorem 6.8 (Wielandt-Yoshida). *Assume* $S \cap G' \ne S \cap (N(S))'$. *Then* $Z_p \wr Z_p$ *is a homomorphic image of* S.

Proof. Let $W = S$ in Theorem 6.7 and apply Lemma 6.4 to $S/\Phi(A)$.

Theorem 6.9 (Wielandt-Yoshida). *Suppose* W *is a weakly closed subgroup of* S *and* $S \cap G' \ne S \cap (N(W))'$. *Then there exist* $x \in S$, $z \in W$, *and a maximal subgroup* W_0 *of* W *such that*

$$[x, z; p - 1] \not\equiv 1 \bmod \Phi(W_0);$$

in particular, $[x, z; p - 1] \ne 1$.

Proof. Apply Theorem 6.7. Since $|\langle W, x \rangle : A| = p$ and $x \in A$, $|W : A \cap W| = p$. Let $W_0 = A \cap W$.

Notes on Chapter I.

§4. Finkel [21] obtained Theorem 4.4 in the case where $\mathfrak{W}$ consists of two conjugacy functors V and W satisfying the following conditions for all p-subgroups P, Q of G:

(a) $V(P) \subseteq Z(P) \subseteq W(P)$ and

(b) if $Z(P) \subseteq Z(Q)$, then $V(P) \subseteq V(Q)$. His methods apply to the factorizations (D1) and (D2) of Theorem B in §II.1.

Theorem 4.6 is taken from [27, Theorem 6.1]. Theorem 4.3 was obtained by the author. We thank G. Seitz and C. R. B. Wright for pointing out a serious error in the original version of Theorem 4.3.

Example 11.3 of GL describes a situation in which $p = 2$ and no set of conjugacy functors on G controls strong fusion or transfer in G. For p odd, we may obtain an example in which no set of conjugacy functors controls strong fusion in G by taking the inverse-transpose automorphism τ of $GL(3, p)$ and assuming that G is the semidirect product of $GL(3, p)$ by $\langle \tau \rangle$. (See Example 11.2 of GL and §III.2 below.)

§5. Theorems 5.5 and 5.10 were obtained by the author.

Corollary 5.7 may be proved without using Corollary 5.6 or other results from §5. One may instead use a lemma:

Assume the hypothesis of Corollary 5.7, and assume that $\Omega_1 Z(S^*) \subseteq W_1(S^*) \cap W_2(S^*)$ *for every* S^*. *Let* B *be a subgroup of* S *that is strongly closed in* S *with respect to at least two of the groups* $C(\Omega_1 Z(S))$, $N(W_1(S))$, $N(W_2(S))$. *Then* B *is strongly closed in* S.

To prove this lemma, assume G is a counterexample of minimal order. For $x, y \in S$, define $x \sim y$ if $x, y \in B$ or $x, y \in S - B$. Then show that $B \cap O_p(G) \lhd G$; $B \cap O_p(G) = 1$; and $\Omega_1 Z(S) \nsubseteq O_p(G)$. This yields a counterexample to Theorem 4.3 in which $\mathfrak{W} = \{\Omega_1 Z, W_1, W_2\}$ and $T \lhd G$.

Chapter II. Factorizations for $p = 2$

1. Introduction. In the previous chapter, we investigated Question 1 of the Preface, which concerns fusion. The other main question of the Preface is the following:

Question 2. What is the relation between S and G if $C(O_p(G)) \subseteq O_p(G)$?

This problem was studied extensively in GL for the case in which p is odd. In this chapter, we prove some results for $p = 2$ and obtain some applications to fusion and simple groups.

It was Thompson who first showed [58] that regardless of p, there is an intimate connection between G and the normalizers of the characteristic subgroups of S if $C(O_p(G)) \subseteq O_p(G)$. By the author's "ZJ-Theorem" (result (14.1) of GL), in many cases G must be equal to the normalizer of some nonidentity characteristic subgroup of S, if p is odd. This result is proved in GL (Theorem 12.3(c)) for a different characteristic subgroup:

(1.1) $K_\infty(S) \triangleleft G$ *if p is odd,* $C(O_p(G)) \subseteq O_p(G)$, *and every section of G is p-stable.* (We note that $K_\infty(S) \neq 1$ if $S \neq 1$.)

For $p = 2$, a reasonable substitute for p-stability is the condition that G be S^4-*free*, i.e., that the symmetric group S^4 not be involved in G (GL, pp. 44–45). Many unsuccessful attempts have been made to find an analogue to (1.1) in this situation. (This problem is discussed as Question III.4.1.) However, in the N-group paper, Thompson obtained a partial analogue (Lemma 5.53) for certain solvable groups, namely, that

(1.2) $$G = N(A)N(B) = N(A)N(C) = N(B)N(C)$$

for some nonidentity characteristic subgroups A, B, C of S. Note that, in the situation of (1.1), one can obtain (1.2) by taking $A = B = C = K_\infty(S)$; in fact, (1.2) yields analogues of many (but not all) of the applications of (1.1).

In later work, Thompson proved [61, p. 2] one of the factorizations in (1.2) for nonsolvable $3'$-groups. This result suggested that (1.2) itself might be generalized by suitably changing the choice of characteristic subgroups. This is done in the following theorem, which is the main result of this chapter. Recall that the groups $Sz(q)$ and $PSL(2, q)$ are defined in §I.2; $J_e(S)$ and $\hat{J}(S)$ will be defined later in this section.

Theorem B. *Suppose $p = 2$. Assume that*
(a) $C(O_2(G)) \subseteq O_2(G)$,
(b) *G is S^4-free, and*

(c) *every non-Abelian composition factor of* G *has the form* $Sz(2^{2n+1})$ *or* $PSL(2, 3^{2n+1})$.

   Then

(D1) $G = C(Z(S))N(J_e(S))$,

(D2) $G = C(Z(S))N(\hat{J}(S))$,

and

(D3) $G = C(\Omega_1 Z\hat{J}(S))N(J_e(S))$.

   The proof of Theorem B will occupy §§3–5 of this chapter. In broad outline, it is similar to the proof of (1.1). Both proofs depend heavily on the examination of the chief factors of G within $O_p(G)$. Such a chief factor, say, M, must be an elementary Abelian p-group. Therefore, M can be regarded as a vector space over the field Z_p, and $G/C_G(M)$ can be regarded as a group of nonsingular linear transformations of M over Z_p. In the proof of (1.1), the hypothesis yields that

   (1.3) $G/C_G(M)$ contains no nonidentity element x for which $(x-1)^2 = 0$.

   This says that the representation of $G/C_G(M)$ on M is a "p-stable" representation in the sense of Gorenstein and Walter [37, p. 103].

   Now suppose that $p = 2$. As in both (1.1) and Theorem B, assume that $C(O_p(G)) \subseteq O_p(G)$. It is easy to see (by Lemma I.2.8) that (1.3) must be violated for some chief factor M within $O_p(G)$, unless $S \lhd G$. Therefore, the condition (1.3) is too restrictive when $p = 2$; in this case, one needs a substitute for (1.3) and the concept of a p-stable representation. It turns out that the conclusion of the following result is an adequate substitute.

   Theorem A. *Suppose* M *is an elementary Abelian 2-group and* G *is a subgroup of* Aut M. *Assume that*

   (a) $O_2(G) = 1$,

   (b) S^3 *is not involved in* G, *and*

   (c) *every non-Abelian composition factor of* G *has the form* $Sz(2^{2n+1})$ *or* $PSL(2, 3^{2n+1})$.

   Then, for every elementary Abelian 2-subgroup A *of* G,

$$|M/C_M(A)| \geq |A|^2 \quad and \quad |[M, A]| \geq |A|^2.$$

   Since this is a rather technical result and its proof does not involve methods used elsewhere in this book, its proof will be given in Appendix A1. However, the interested reader may obtain some feeling for the proof by verifying the special case in which G is a dihedral group and $|G|/2$ is odd.

   As mentioned in the Preface, most of the ideas used in proving Theorems A and B remain evident (or become more evident) if the reader assumes throughout that G is solvable.

   Now let us assume that G and p are arbitrary. By applying the reductions in Chapter I to (1.1), one can obtain the following result (Theorem 12.9 of GL):

(1.4) K_∞ controls strong fusion in G if p is odd and every section of G is p-stable.

In $\S 6$, we apply similar reductions to Theorem B and obtain a weak analogue of (1.4).

An important result of David Goldschmidt classifies all the non-Abelian simple groups in which a Sylow 2-subgroup has a nonidentity strongly closed Abelian subgroup. We will call these groups the *Goldschmidt groups*. As a corollary of this theorem and the results of $\S 6$, we will obtain the following result (and various applications) in $\S 7$.

Theorem C. *Suppose G is a simple non-Abelian group. Then the following conditions are equivalent:*

(a) *G is S^4-free;*

(b) *G is a Goldschmidt group.*

Now we introduce the special notation for this chapter.

For every subset A in G, let A^G be the set-theoretic union

$$A^G = \bigcup_{x \in G} A^x.$$

(This differs from the usage in [37].)

Definitions. Suppose T is a 2-group. Let

$d_e(T) = \max\{|A| \,|\, A$ is an elementary Abelian subgroup of $T\}$,

$\mathcal{Q}_e(T) = \{A \,|\, A$ is an elementary Abelian subgroup of T and $|A| = d_e(T)\}$,

$J_e(T) = \langle \mathcal{Q}_e(T) \rangle$, $\Omega_1 Z J_e(T) = \Omega_1 Z(J_e(T))$.

We say that T is an *E-group* if $Z(T)$ contains every normal elementary Abelian subgroup V of T that has the following property:

Whenever R is a nonidentity elementary Abelian subgroup of $T/C_T(V)$, then

$$|V/C_V(R)| > |R|^{3/2} \quad and \quad |[V, R]| > |R|.$$

Definition. Let T be a 2-group. Then

$\hat{J}(T) = \langle T^* \,|\, T^*$ is an E-group and $J_e(T) \subseteq T^* \subseteq T \rangle$ and $\Omega_1 Z \hat{J}(T) = \Omega_1 Z(\hat{J}(T))$.

Remark 1.1. Let T be a 2-group. Lemma 2.1 (in $\S 2$) and its proof show that $J_e(T)$ is an E-group and $J_e(T) \subseteq \hat{J}(T)$.

Remark 1.2. In [29], we have proved the conclusion of Theorem B under a much more general hypothesis. (See Example 8.4 and the main result in [29].) To a large extent, one can use conditions on chief factors weaker than those in the conclusion of Theorem A, e.g., $|M/C_M(A)| > |A|^{3/2}$ and $|[M, A]| > |A|$.

Remark 1.3. Many of the familiar linear groups over finite fields of characteristic 2 violate the conclusion of Theorem A and even the weaker condition in Remark 1.2 (Example 8.5 of [29]). This condition may be violated in a different way by allowing A to be a group which is Abelian but not elementary, e.g., when $|M| = 2^4$, $|A| = 4$, and G is the Frobenius group of order 20 ('Sz(2)').

2. Preliminary lemmas. For this chapter, we require a number of particular results about the J-subgroups and groups in which S^3 is not involved.

Lemma 2.1. *Let* P *be a finite* p-*group.*

(a) *For every* $A \in \mathcal{Q}_e(P)$,

$$\Omega_1(C_P(J_e(P))) = \Omega_1 Z J_e(P) \subseteq A = \Omega_1(C_P(A)).$$

(b) *If* $J_e(P) \subseteq Q \subseteq P$, *then* $J_e(P) = J_e(Q)$.

(c) *For* $J = J_e(P)$, *we have* $J_e(P) = J_e(\hat{J}(P)) \subseteq \hat{J}(P)$ *and* $\Omega_1 Z \hat{J}(P) = \Omega_1(C_J(\hat{J}(P)))$.

(d) *If* $\hat{J}(P) \subseteq Q \subseteq P$, *then* $\hat{J}(P) = \hat{J}(Q)$.

(e) *The subgroup* $\Omega_1 Z J_e(P)$ *contains every normal elementary Abelian subgroup* V *of* P *with the following property*:

Whenever B *is a nonidentity elementary Abelian subgroup of* $P/C_P(V)$, *then* $|V/C_V(B)| > |B|^{3/2}$ *and* $|[V, B]| > |B|$.

Proof. Suppose $A \in \mathcal{Q}_e(P)$. If $x \in C_P(A)$ and x has order p, then $\langle A, x \rangle$ is elementary Abelian, so $|\langle A, x \rangle| \leq |A|$. This shows that

$$A = \Omega_1(C_P(A)) \supseteq \Omega_1(C_P(J_e(P))),$$

which yields (a).

Now we prove (e). Assume it is violated for some V. By (a), $V \not\subseteq \Omega_1(C_P(J_e(P)))$. Therefore, there exists $A \in \mathcal{Q}_e(P)$ such that A does not centralize V. Let $B = A C_P(V)/C_P(V)$. Then $V C_A(V)$ is elementary Abelian and $C_V(A) = C_V(B)$. Since $A \in \mathcal{Q}_e(P)$,

$$|A| \geq |C_A(V)V| = |C_A(V)||V/(V \cap C_A(V))| \geq |C_A(V)||V/C_V(A)|.$$

Hence, $|A/C_A(V)| \geq |V/C_V(A)| = |V/C_V(B)|$. By the given property of V, $|V/C_V(B)| > |B|^{3/2} = |A/C_A(V)|^{3/2}$, a contradiction. This proves (e).

In (b), $d_e(P) = d_e(Q)$ and we easily see that $\mathcal{Q}_e(P) = \mathcal{Q}_e(Q)$ and $J_e(P) = J_e(Q)$.

By (b), $J_e(P) = J_e(J_e(P))$. By (e) and the definition of an E-group, $J_e(P)$ is an E-group. Hence,

$$\hat{J}(P) = \langle P^* | J_e(P) \subseteq P^* \subseteq P \text{ and } P^* \text{ is an } E\text{-group} \rangle \supseteq J_e(P).$$

By (b), $J_e(P) = J_e(\hat{J}(P))$. Therefore, by (a),

$$\Omega_1(C_P(\hat{J}(P))) \subseteq \Omega_1 Z J_e(P) \subseteq J_e(P) \subseteq \hat{J}(P),$$

from which (c) follows.

Suppose $\hat{J}(P) \subseteq Q \subseteq P$. By (c) and (b), $J_e(P) = J_e(Q)$. Now the proof of (d) is similar to that of (b). This completes the proof of Lemma 2.1.

Lemma 2.2. *Suppose* p *is an odd prime. The following conditions are equivalent*:

(a) *the dihedral group of order* $2p$ *is involved in* G;

(b) *some* 2-*element of* G *normalizes but does not centralize some* p-*subgroup of* G;

(c) *some* 2-*element of* G *inverts some nonidentity* p-*element of* G.

Proof. It is easy to show that (c) yields (a). We will prove that (a) yields (b) and that (b) yields (c).

Assume (a). Take H, $K \subseteq G$ such that $K \lhd H$ and H/K is dihedral of order $2p$. Let T be a Sylow p-subgroup of $H'K$. By Lemma I.2.2 (c), $H'K/K = TK/K$ and $H'K = TK$. By the Frattini argument,

$$H = N_H(T)H'K = N_H(T)TK = N_H(T)K.$$

Therefore, $N_H(T)/N_K(T) \cong H/K$. Consequently, every 2-element in $N_H(T)$ that lies outside $N_K(T)$ will invert $TN_K(T)/N_K(T)$ and hence will not centralize T. This proves (b).

Now assume (b). Take a p-subgroup U of G and a 2-element x of G such that $x \in N(U) - C(U)$. We may choose x such that x^2 centralizes U. Choose $y \in U - C_U(x)$; then $y^{-1}y^x \neq 1$ and x inverts $y^{-1}y^x$. This proves (c) and completes the proof of the lemma.

Lemma 2.3. *The following conditions are equivalent:*
(a) S^4 *is involved in* G;
(b) *there exists a 2-subgroup T of G such that S^3 is involved in* $N(T)/C(T)$.

Proof. First, assume (a). Take subgroups H and K of G such that $K \lhd H$ and $H/K \cong S^4$. By Lemma I.2.4, there exists $L \subseteq G$ such that

$$LK = H \quad \text{and} \quad O_2(H/K) = O_2(L)K/K \cong O_2(L)/(O_2(L) \cap K).$$

Then a section of L isomorphic to S^3 acts faithfully on $O_2(L)/(O_2(L) \cap K)$. Therefore S^3 is involved in $N(O_2(L))/C(O_2(L))$.

Now assume (b). We prove (a) by induction on $|G|$. Therefore, we may assume that

(2.1)   no section of G except G itself satisfies (b).

By (2.1), $N(T) = G$ and $T \subseteq O_2(G)$. Therefore, S^3 is involved in $G/C(T)$ and hence in $G/C(O_2(G))$. Replacing T by $O_2(G)$ if necessary, we may assume that

(2.2) $T = O_2(G).$

Let $C = C(T)$. Let q be any odd prime and let Q be a Sylow q-subgroup of C. By the Frattini argument, $G = CN(Q)$. Therefore, $N(Q)$ and G induce the same group of automorphisms on T by conjugation. By (2.1), $G = N(Q)$. Similarly, by considering G/Q and TQ/Q in place of G and T, we see that $Q = 1$. Since q was chosen arbitrarily, C is a 2-group. As $C \lhd G$,

(2.3) $C \subseteq O_2(G) = T.$

Let $C_0 = C_G(T/\Phi(T))$. By (b) and Lemma 2.2, there exist a 2-element x and a 3-element y in G such that x inverts y. We may choose y of order 3. By (2.3), $y \notin C$. By Lemma I.2.6 (d), $y \notin C_0$. Therefore, the coset yC_0 is an element of G/C_0

of order 3 which is inverted by xC_0. By Lemma 2.2, S^3 is involved in $\langle x, y, C_0 \rangle / C_0$. By two successive applications of (2.1),

(2.4) $G = \langle x, y, T \rangle$ and $\Phi(T) = 1.$

By (2.4) and Lemma I.2.6, T is elementary Abelian and so

(2.5) $C_T(y) \lhd \langle T, y, x \rangle = G.$

By Lemma I.2.7, y acts nontrivially on $T/C_T(Y)$. Consequently, by (2.1),

(2.6) $C_T(y) = 1.$

Since x^2 centralizes y,

$$\langle x^2, T \rangle \lhd G, \quad \langle x^2, T \rangle \subseteq O_2(G) = T, \quad \text{and} \quad x^2 \in C_T(y) = 1.$$

Therefore, $\langle x, y \rangle \cong S^3$. Take $z \in C_T(x)^{\#}$; then, by (2.4),

(2.7) $\langle z^G \rangle = \langle z^{\langle x, y \rangle} \rangle = \langle z, z^y, z^{y^2} \rangle$

and $|\langle z^G \rangle| \le 8$. By (2.1) and (2.2), $G = \langle x, y, z^G \rangle$. Since $|G : N(\langle y \rangle)| \equiv 1 \pmod 3$, by Sylow's Theorem, a short argument yields a permutation representation of G of degree 4, which then yields a homomorphism of G onto S^4. This completes the proof of Lemma 2.3.

Lemma 2.4. *Suppose* $C(O_2(G)) \subseteq O_2(G)$ *and* H *is a subgroup of* G. *Assume that* H *or* $O^2(H)$ *is a normal subgroup of* G. *Then* $C_H(O_2(H)) \subseteq O_2(H)$.

Proof. Let $T = O_2(G)$.

Assume first that $H \lhd G$. Then $[T, H] \subseteq T \cap H \subseteq O_2(H)$. Consequently, $C_H(O_2(H))$ stabilizes the normal series $T \supseteq O_2(H) \supseteq 1$ of T. By Lemma I.2.8, $C_H(O_2(H))$ induces a 2-group of automorphisms of T. Since $C_H(O_2(H)) \lhd G$ and

$$O_2(G/C_G(T)) = O_2(G/Z(T)) = T/Z(T)$$

we have $C_H(O_2(H)) \subseteq T \cap H \subseteq O_2(H)$, as desired.

Now assume that $O^2(H)$ is a normal subgroup of G. Let x be an arbitrary element of odd order in $C_H(O_2(H))$. Since $O_2(O^2(H)) \subseteq O_2(H)$, $x \in C_{O^2(H)}(O_2(O^2(H)))$. Then the previous paragraph yields that $x = 1$. Thus, $C_H(O_2(H))$ is a 2-group. Therefore, $C_H(O_2(H)) \subseteq O_2(H)$, as desired.

Lemma 2.5. *Suppose* A *is an operator group on* G *and* $G = G_0 \supseteq G_1 \supseteq \cdots \supseteq G_n = 1$ *is a series of subgroups of* G *fixed by* A. *Assume that* $G_i \lhd G_{i-1}$ *for* $i = 1, \ldots, n$. *Then*

$$|G : C(A)| \ge \prod_{1 \le i \le n} |G_{i-1}/G_i : C_{G_{i-1}/G_i}(A)|.$$

Proof. We use induction on n. If $n = 1$, then $G_1 = 1$ and the result is trivial. Assume $n > 1$.

Let $N = G_1$ and let

$$r = \prod_{2 \leq i \leq n} |G_{i-1}/G_i : C_{G_{i-1}/G_i}(A)|.$$

By induction, $r \leq |N:C_N(A)|$. Now

$$|C_G(A):C_N(A)| = |C_G(A)N/N| \leq |C_{G/N}(A)|.$$

Therefore,

$$|G:C_G(A)|/r \geq |G/N|/|C_G(A):C_N(A)| \geq |G/N|/|C_{G/N}(A)|$$

$$= |G/N:C_{G/N}(A)|.$$

Hence, $|G:C(A)| \geq |G/N:C_{G/N}(A)|r$, which immediately yields the conclusion.

Lemma 2.6. *Suppose n is an odd natural number.*

(a) *Each nonsolvable subgroup of* $\mathrm{PSL}(2, 3^n)$ *is isomorphic to* $\mathrm{PSL}(2, 3^m)$ *for some odd m.*

(b) *The Sylow 2-subgroups of* $\mathrm{Sz}(2^n)$ *are trivial intersection sets.*

(c) *Each nonsolvable subgroup of* $\mathrm{Sz}(2^n)$ *is isomorphic to* $\mathrm{Sz}(2^m)$ *for some m.*

(d) *The factor group* $(\mathrm{Aut}\, \mathrm{Sz}(2^n))/(\mathrm{In}\, \mathrm{Sz}(2^n))$ *has order n.*

(e) *For each involution x of* $\mathrm{Sz}(2^n)$ *and each odd prime divisor t of* $|\mathrm{Sz}(2^n)|$, *there exists a cyclic Sylow t-subgroup of* $\mathrm{Sz}(2^n)$ *inverted by x.*

(f) *The group S^3 is neither involved in* $\mathrm{PSL}(2, 3^n)$ *nor in* $\mathrm{Sz}(2^n)$.

Proof. Part (a) follows from a theorem of Dickson [44, pp. 213–214]; in fact, m must divide n in each case. Parts (b), (c), and (d) follow from results of Suzuki about the groups $\mathrm{Sz}(q)$ [57, pp. 137–139]. These results also yield in (e) that $\mathrm{Sz}(2^n)$ has a cyclic Sylow t-subgroup contained in a dihedral subgroup; since all involutions in $\mathrm{Sz}(2^n)$ are conjugate, we obtain (e). Since $\mathrm{Sz}(2^n)$ is a $3'$-group, it does not involve S^3.

Suppose $G = \mathrm{PSL}(2, 3^n)$ and S^3 is involved in G. By Lemma 2.2, some 2-element x of G inverts some nonidentity 3-element y of G. Let T be a Sylow 3-subgroup of G that contains y. By [44, p. 191], T is a trivial intersection set and G has $3^n + 1$ Sylow p-subgroups. Therefore, $x \in N(T)$ and

$$|N(T)| = |G|/(3^n + 1) = 3^n(3^n - 1)/2.$$

As n is odd, $3^n - 1 \equiv 2 \pmod 4$. Hence $|N(T)|$ is odd. This is impossible because $x \in N(T)$. This contradiction completes the proof of (f) and of the lemma.

Part (f) of Lemma 2.6 shows that both cases (a) and (b) of the next lemma are possible. Later (in §7) we shall see that these include all non-Abelian simple groups in which S^3 is not involved.

Lemma 2.7. *Suppose H is a minimal normal subgroup of G and n is a natural number. Assume that S^3 is not involved in G, that $O_2(G) = 1$, and that $H = O^2(G)$. Assume further that H is the direct product of isomorphic simple groups $J_1, \ldots, J_k$ of the form* $\mathrm{Sz}(2^{2n+1})$ *or* $\mathrm{PSL}(2, 3^{2n+1})$. *Then:*

(a) if $J_1 \cong Sz(2^{2n+1})$, then $H = \bigcap_{1 \le i \le k} N(J_i)$ and

(b) if $J_1 \cong PSL(2, 3^{2n+1})$, then $G = H = J_1$.

Proof. Clearly, $C(H) \cap H = 1$. Since G/H is a 2-group, $C(H)$ is a 2-group and

(2.8) $C(H) \subseteq O_2(G) = 1$.

(a) Assume $J_1 \cong Sz(2^{2n+1})$. Let $K = \bigcap_{1 \le i \le k} N(J_i)$. Then $H \subseteq K \subseteq G$, whence K/H is a 2-group. By Lemma 2.6 (d),

$$|\text{Aut } Sz(2^{2n+1})/\text{In } Sz(2^{2n+1})| = 2n + 1,$$

which is odd. Therefore, every element of K induces an inner automorphism on H by conjugation. By (2.8), $K \subseteq HC(H) = H$. Thus $H = K$, as claimed.

(b) Assume $J_1 \cong PSL(2, 3^{2n+1})$. Let T be a Sylow 3-subgroup of H. By the Frattini argument,

(2.9) $G = HN(T)$.

Since S^3 is not involved in G, every 2-element of $N(T)$ centralizes T, by Lemma 2.2. By hypothesis, G/H is a 2-group. Therefore $N(T)/N_H(T)$ is a 2-group and $N(T) = N_H(T)C(T)$. Hence, by (2.9),

(2.10) $G = HC(T)$ and $C(T)/C_H(T)$ is a 2-group.

Since $T \cap H = (T \cap J_1) \times \cdots \times (T \cap J_k)$ and $T \cap J_i$ is a Sylow 3-subgroup of J_i for each i, $C(T)$ normalizes $J_1, \ldots, J_k$. As H also normalizes $J_1, \ldots, J_k$ and H is a minimal normal subgroup of G, (2.10) yields that

(2.11) $H = J_1 \cong PSL(2, 3^{2n+1})$.

The elements of G act by conjugation on the family $\mathcal{F}$ of all Sylow 3-subgroups of J_1. Let L be the kernel of this representation. It is easy to see that H acts faithfully and transitively on $\mathcal{F}$ and that

(2.12) T acts transitively and regularly on the set $\mathcal{F} - \{T\}$.

Therefore, $[H, L] \subseteq H \cap L = 1$ and, by (2.8), $1 = C(H) \supseteq L$. Thus, G acts faithfully on $\mathcal{F}$. Since T is Abelian, (2.12) yields that $C(T) = T$, by a short argument on permutation groups. Since $T \subseteq H$, (2.10) and (2.11) yield that $G = H = J_1$. This completes the proof of (b) and of Lemma 2.7.

Lemma 2.8. *Suppose V is a nonidentity normal elementary Abelian subgroup of S and H is a minimal normal subgroup of G. Assume that*

(a) $p = 2$ and S^3 is not involved in G,

(b) $O_2(G) = 1$,

(c) $H = O^2(G)$,

(d) every non-Abelian composition factor of G has the form $Sz(2^{2n+1})$ or $PSL(2, 3^{2n+1})$, and

(e) *whenever $V \subseteq F \subset G$ and $F \cap S$ is a Sylow 2-subgroup of F, then $V \subseteq O_2(F)$.*
Then there exists $x \in G$ such that $\langle V, V^x \rangle = G$.

Proof. By (c), G/H is a 2-group. Hence, by Lemma I.2.2 (c), $G = HS$. Then $C_S(H) \triangleleft HS = G$ and

(2.13) $$C_S(H) \subseteq O_2(G) = 1.$$

Now,

(2.14) H is either elementary Abelian of odd order or a direct product of isomorphic simple groups transitively permuted under conjugation by G.

Therefore,

$$[V \cap O_2(HV), H] \subseteq O_2(HV) \cap H \subseteq O_2(H) = 1.$$

By (2.13), $V \cap O_2(HV) \subseteq C_S(H) = 1$. As $HV \triangleleft HS = G$, condition (e) yields that

(2.15) $$G = HV.$$

Suppose H is Abelian. By (2.14) and (2.15), V is a Sylow 2-subgroup of G. By (b), $V \ntriangleleft G$. Take $x \in G$ such that $V^x \neq V$. Let $F = \langle V, V^x \rangle$. If $V \subseteq O_2(F)$, then $V = O_2(F)$, $|F/V|$ is odd, and $V^x = V$, a contradiction. Therefore, $V \nsubseteq O_2(F)$. By (e), $G = F = \langle V, V^x \rangle$, as desired.

Suppose H is not Abelian. By (a)–(d) and (2.14), the hypothesis of Lemma 2.7 is satisfied for some natural numbers n and k and some simple groups $J_1, \ldots, J_k$. If $G \cong PSL(2, 3^{2n+1})$, then for any $x \in S^{\#}$, the group $C(x)$ is dihedral of order $3^{2n+1} + 1$, and (e) is violated with $F = C(x)$ for some x. Hence, $G \ncong PSL(2, 3^{2n+1})$. By Lemma 2.7,

(2.16) $$H = \bigcap_{1 \leq i \leq k} N(J_i) \quad \text{and} \quad J_1 \cong Sz(2^{2n+1}).$$

Suppose $k \geq 2$. Now, $S \cap H$ is a Sylow 2-subgroup of H and $S \cap J_i$ is a Sylow 2-subgroup of J_i for each i. By (2.14) and (2.15), there exists $x \in V$ such that $J_1^x = J_2$. Take $y \in S \cap J_1$ of order four. Then $y^x \in S \cap J_2$. Since $\langle S \cap J_1, S \cap J_2 \rangle = (S \cap J_1) \times (S \cap J_2)$, $y^{-1}y^x$ has order four. Since $V \triangleleft S$, $y^{-1}y^x = (y^{-1}x^{-1}y)x \in V$, contrary to the fact that V is elementary Abelian. Thus, $k = 1$ and $J_1 = H \triangleleft G$. By (2.16), $G = H \cong Sz(2^{2n+1})$.

Take any $x \in G - N(S)$. Let $F = \langle V, V^x \rangle$ and let R be a Sylow 2-subgroup of F that contains V. By Lemma 2.6(b), the Sylow 2-subgroups of G are trivial intersection subgroups, $R \subseteq S$, and $1 = V^x \cap R \supseteq V^x \cap O_2(F)$. Consequently, $V \nsubseteq O_2(F)$. By (e), $G = F = \langle V, V^x \rangle$. This completes the proof of Lemma 2.8.

3. Proof of (D1) and (D2). In this section, we prove the first two factorizations of Theorem B, namely,

(D1) $$G = C(Z(S))N(J_e(S))$$

and

(D2) $G = C(Z(S))N(\hat{J}(S))$.

By Lemma 2.1, $J_e(S) = J_e(\hat{J}(S))$. Therefore, $N(\hat{J}(S)) \subseteq N(J_e(S))$. Thus, (D1) follows from (D2), and it will suffice to prove (D2).

Let $T = O_2(G)$ and let C be the largest normal subgroup of G that centralizes $Z(S)$. Let $C_0/C = O_2(G/C)$. Then $C_0 \triangleleft G$ and, by Lemma I.2.2 (d), $C_0 \subseteq CS \subseteq C(Z(S))$. By our choice of C, we have $C_0 = C$ and

(3.1) $O_2(G/C) = 1$.

By condition (a) of the hypothesis, $T \supseteq C(T) \supseteq C(S) \supseteq Z(S)$. Since $T \subseteq S$, $T \subseteq C$. Therefore,

(3.2) $Z(S) \subseteq Z(C)$ and $C = C(Z(C))$.

Moreover,

(3.3) $Z(C) \subseteq C(T) = Z(T)$.

Let $V = \Omega_1(Z(C))$. Since $Z(C)$ is Abelian, the automorphisms of $Z(C)$ that act trivially on V form a 2-group, by Lemma I.2.7. Therefore, $C(V)/C(Z(C))$ is a 2-group and thus is a normal 2-subgroup of $G/C(Z(C))$. By (3.2) and (3.1), $C(Z(C)) = C$ and $C(V)/C \subseteq O_2(G/C) = 1$, which yields that

(3.4) $C(V) = C$ and $O_2(G/C(V)) = 1$.

By Lemma 2.3 and conditions (a) and (b) of the hypothesis, $C(T) = Z(T)$ and S^3 is not involved in $G/Z(T)$. By (3.2) and (3.3), $C = C(Z(C)) \supseteq Z(T)$. By (3.4), $C = C(V)$. Therefore,

(3.5) S^3 is not involved in $G/C(V)$.

By condition (c) of the hypothesis, every non-Abelian composition factor of G, and hence of G/C, has the form $Sz(2^{2n+1})$ or $PSL(2, 3^{2n+1})$. Consequently, by (3.4), (3.5), and Theorem A,

(3.6) for every nonidentity elementary Abelian subgroup B of SC/C,

$$|V/C_V(B)| \geq |B|^2 > |B|^{3/2}   \text{ and } |[V, B]| \geq |B|^2 > |B|.$$

Suppose R is an E-group and $J_e(S) \subseteq R \subseteq S$. By (3.6) and Lemma 2.1, $V \subseteq \Omega_1 Z J_e(S) \subseteq J_e(S) \subseteq R$. Since $V \triangleleft G$, $V \triangleleft R$. By (3.6) and the definition of an E-group, $Z(R) \supseteq V$. By (3.4), $C = C(V) \supseteq R$. Since R was chosen arbitrarily, it follows that $\hat{J}(S) = \langle S^* | S^* \text{ is an } E\text{-group and } J_e(S) \subseteq S^* \subseteq S \rangle \subseteq C$.

Now, $\hat{J}(S) \subseteq S \cap C$. By Lemma 2.1,

(3.7) $\hat{J}(S) = \hat{J}(S \cap C)$.

By the Frattini argument, $G = CN(S \cap C)$. By definition, C centralizes $Z(S)$. By (3.7),

$$N(S \cap C) \subseteq N(\hat{J}(S \cap C)) = N(\hat{J}(S)).$$

Consequently, $G = C(Z(S))N(\hat{J}(S))$. This proves (D2). As mentioned above, (D1) follows from (D2).

4. **The counterexample to (D3).** In the previous section, we proved the factorizations (D1) and (D2). To complete the proof of Theorem B, we must obtain (D3). Therefore, we now assume that G and S violate (D3) and that the order of G is minimal subject to this condition. Thus, $G \neq C(\Omega_1 Z \hat{J}(S))N(J_e(S))$. We will reach a contradiction at the end of the next section.

Let $J = J_e(S)$, $T = O_2(G)$, $Z = \Omega_1 Z \hat{J}(S)$, and $Z^* = \langle Z^{N(J)} \rangle$. Choose L to be the largest normal subgroup of G that normalizes Z^*.

Proposition 4.1. *There exists a normal elementary Abelian subgroup V of S such that $V \not\subseteq Z(S)$ and V has the following property:*

Whenever B is a subgroup of S and $B/C_B(V)$ is a nonidentity elementary Abelian group, then

$$|V/C_V(B)| > |B/C_B(V)|^{3/2} \quad and \quad |[V, B]| > |B/C_B(V)|.$$

Proof. Assume no such subgroup V exists. By the definition of an E-group and a short argument, S is an E-group. Obviously, $S \supseteq J_e(S)$. By the definition of $\hat{J}(S)$, $\hat{J}(S) = \langle S^* | S^* $ is an E-group and $J_e(S) \subseteq S^* \subseteq S \rangle = S$.

In the previous section, we proved the factorization (D1), namely $G = C(Z(S))N(J)$. Hence

$$G = C(\Omega_1 Z(S))N(J) = C(\Omega_1 Z \hat{J}(S))N(J).$$

But this contradicts the choice of G.

Henceforth, we take V to be a fixed subgroup of S that satisfies the conditions of Proposition 4.1. In this section, we will prove that $V \subseteq T$.

Proposition 4.2. *The following conditions are satisfied:*
(a) $Z \subseteq Z^* \subseteq \Omega_1 Z(J)$;
(b) $V \subseteq \Omega_1 Z(J)$;
(c) *whenever U is a subgroup of S that contains V, then $V \subseteq \Omega_1 Z J_e(U)$;*
(d) *whenever F is a subgroup of G, then every non-Abelian composition factor of F has the form $Sz(2^{2n+1})$ or $PSL(2, 3^{2n+1})$.*

Proof. By Lemma 2.1 (c), $Z \subseteq \Omega_1 Z(J)$. Hence

$$Z \subseteq \langle Z^x | x \in N(J) \rangle \subseteq \langle (\Omega_1 Z(J))^{N(J)} \rangle \subseteq \Omega_1 Z(J),$$

which yields (a). Parts (b) and (c) follow from Lemma 2.1 (e) and the choice of V. For (d), we note that, as in any subgroup of a group, the composition factors of F are isomorphic to sections of the composition factors of G. Therefore, (d) follows from Lemma 2.6.

Lemma 4.3. *Suppose F is a proper subgroup of G that contains J. Assume that $C_F(O_2(F)) \subseteq O_2(F)$ and that $F \cap S$ is a Sylow 2-subgroup of F. Then*

(a) $F = C_F(Z)N_F(J)$,

(b) $[F, J] \subseteq C(Z)$, and

(c) $FL \subset G$.

Proof. Since $J_e(S) = J \subseteq F \cap S$,

(4.1) $$J = J_e(F \cap S).$$

Consequently,

$$\hat{J}(F \cap S) = \langle S^* | J_e(F \cap S) \subseteq S^* \subseteq F \cap S \text{ and } S^* \text{ is an } E\text{-group} \rangle$$

$$\subseteq \langle S^* | J_e(S) \subseteq S^* \subseteq S \text{ and } S^* \text{ is an } E\text{-group} \rangle = \hat{J}(S).$$

Therefore, by (4.1) and Lemma 2.1 (c),

(4.2) $$Z = \Omega_1 Z \hat{J}(S) = \Omega_1(C_J(\hat{J}(S))) \subseteq \Omega_1(C_J(\hat{J}(F \cap S))) = \Omega_1 Z \hat{J}(F \cap S).$$

By Proposition 4.2 (d) and the hypothesis of this lemma, the hypothesis of Theorem B is satisfied by F and $F \cap S$ in place of G and S. Hence, by (4.1) and induction,

$$F = C_F(\Omega_1 Z \hat{J}(F \cap S))N_F(J).$$

By (4.2), $F = C_F(Z)N_F(J)$, which yields (a).

Let $Y = \langle Z^F \rangle$. Then $Y \lhd F$ and

(4.3) $$C_F(Y) \lhd F.$$

By (a) and Proposition 4.2 (a),

(4.4) $$Y = \langle Z^F \rangle = \langle Z^{N_F(J)} \rangle \subseteq \langle Z^{N_G(J)} \rangle = Z^* \subseteq \Omega_1 Z(J).$$

Hence $J \subseteq C_F(Y)$. By (4.3),

$$[F, J] \subseteq [F, C_F(Y)] \subseteq C_F(Y) \subseteq C_F(Z) \subseteq C_G(Z).$$

This proves (b).

Suppose $FL = G$. By (4.4) and the definition of Z^* and L,

$$Z^* = \langle Z^{N(J)} \rangle \subseteq \langle Z^G \rangle = \langle Z^{xy} | x \in F, y \in L \rangle$$

$$\subseteq \langle (Z^*)^y | y \in L \rangle = Z^*.$$

Hence, $Z^* = \langle Z^G \rangle \lhd G$. By the Frattini argument,

(4.5) $$G = C(Z^*)N(C(Z^*) \cap S) = C(Z^*)N(C_S(Z^*)).$$

By (4.4), $Z^* \subseteq \Omega_1 Z(J)$. Therefore, $J \subseteq C_S(Z^*)$ and, by Lemma 2.1, $J = J_e(C_S(Z^*)) \lhd N(C_S(Z^*))$. By (4.4), $Z \subseteq Z^*$. Consequently, by (4.5), $G = C(Z)N(J)$. But this contradicts the choice of G. Thus, $FL \neq G$, which completes the proof of (c) and of Lemma 4.3.

By Lemma 4.3 (c) with $F = S$, $L \subseteq LS \subset G$. Let H/L be a minimal normal subgroup of G/L.

Corollary 4.4. *The group G satisfies the following conditions:*
(a) $G = \langle J^G \rangle = 0^2(G)J$;
(b) L *contains every normal subgroup* N *of* G *for which* $NJ \subset G$;
(c) $HJ = G$;
(d) $C_S(H/L) = S \cap L$;
(e) L *contains* T *and* L/T *has odd order*;
(f) $L/T \subseteq \Phi(G/T)$;
(g) *for every noncentral chief factor* F *of* G *within* T, $C_S(F) = T$;
(h) $0_2(G/L) = 1$, $H/L = 0^2(G/L)$, *and* S^3 *is not involved in* G/L.

Proof. (a) Assume $\langle J^G \rangle \subset G$. Let $F = \langle J^G \rangle$. Then $F \supseteq J$ and $F \lhd G$. Hence, $F \cap S$ is a Sylow 2-subgroup of F. Since $C(0_2(G)) \subseteq 0_2(G)$ and $F \lhd G$, Lemma 2.4 yields that $C_F(0_2(F)) \subseteq 0_2(F)$. By Lemma 4.3,

$$(4.6) \qquad F = C_F(Z)N_F(J).$$

Since $J \subseteq F$, we have $J \subseteq F \cap S$. By Lemma 2.1, $J = J_e(F \cap S)$. Therefore, by the Frattini argument, $G = FN(F \cap S) = FN(J)$. By (4.6),

$$G = (C_F(Z)N_F(J))N(J) = C_F(Z)N(J) = C(Z)N(J),$$

contrary to the choice of G. Thus, $\langle J^G \rangle = G$.

Since $G/0^2(G)$ is a 2-group, $G = S0^2(G)$ and $G = \langle J^G \rangle = \langle J^{S0^2(G)} \rangle = \langle J^{0^2(G)} \rangle \subseteq J0^2(G)$. Hence $G = J0^2(G)$.

(b) Suppose $N \lhd G$ and $NJ \subset G$. Let $F = NJ$. Then $J \subseteq F$. By Lemma I.2.2(e), $F \cap S$ is a Sylow 2-subgroup of F. Since $N \lhd G$ and F/N is a 2-group, Lemma 2.4 yields that $C_F(0_2(F)) \subseteq 0_2(F)$. Therefore,

$$(4.7) \qquad F = C_F(Z)N_F(J)$$

by Lemma 4.3.

Let $K = N(J)$. Then K normalizes N and F, and, by (4.7),

$$FK = (C_F(Z)N_F(J))K = C_F(Z)K.$$

Hence, by the definition of Z^*, $\langle Z^{FK} \rangle = \langle Z^K \rangle = Z^*$. Since $\langle Z^{FK} \rangle \lhd FK$, $N \subseteq FK \subseteq N(Z^*)$. Now, $N \subseteq L$ by the definition of L.

(c) This follows from (b) and the definition of H.

(d) Since $C_{LS}(H/L)$ contains L, both H and S normalize $C_{LS}(H/L)$. By (c), $G = HJ = HS$. Therefore,

$$(4.8) \qquad C_{LS}(H/L) \lhd G.$$

The definitions of L and Z^* yield that

$$C_{LS}(H/L) = LS \cap C_G(H/L) = L(S \cap C_G(H/L)) \subseteq N(Z^*).$$

Because of (4.8), the definition of L yields that $L \supseteq C_{LS}(H/L)$. Hence, $C_S(HL/L) = L \cap S$, as desired.

(e) Let $F = N(L \cap S)$. Then $\langle J, T \rangle \subseteq S \subseteq F$ and $C_F(O_2(F)) \subseteq C_F(T) \subseteq C_G(T) \subseteq T \subseteq O_2(F)$. From the Frattini argument, $G = FL$. Consequently, by Lemma 4.3, $F = G$. Thus,

$$(4.9) \qquad\qquad L \cap S \lhd G.$$

By (b), $T \subseteq L$. Then $T \subseteq L \cap S$. However, (4.9) yields that $L \cap S \subseteq O_2(G) = T$. Therefore, $L \cap S = T = L \cap S$ and $|L/T|$ is odd, as desired.

(f) Suppose this is false. Then there exists a maximal subgroup F/T of G/T such that $L/T \not\subseteq F/T$. Then $G/T = (F/T)(L/T)$, whence $G = FL$. By (e), F/T contains a Sylow 2-subgroup of G/T. Replacing F by a conjugate if necessary, we may assume that $F \supseteq S$. Since $T \subseteq O_2(F)$, we have $C_F(O_2(F)) \subseteq O_2(F)$. Now Lemma 4.3 yields that $G \neq FL$, a contradiction.

(g) Suppose F is a noncentral chief factor of G within T and $C = C(F)$. Then $G \supset C$. By Lemma I.2.2(f), $T \subseteq C$ and $O_2(G/C) = 1$. Therefore, $T \subseteq C \cap S$ and $G \supset CS \supseteq CJ$. By (b), $C \subseteq L$. Hence, (e) yields that

$$C_S(F) = C \cap S \subseteq L \cap S = T \cap S \subseteq T \subseteq C \cap S.$$

Thus, $C_S(F) = C \cap S = T$.

(h) By the hypothesis of Theorem B and Lemma 2.3, S^3 is not involved in G/L.

Because of our choice of H, H/L is a direct product of simple groups. By (d), $C_{SL/L}(H/L) = 1$. Therefore, H/L is not a 2-group and $O^2(G/L) = O^2(HJ/L) = H/L$. Moreover,

$$[O_2(G/L), H/L] \subseteq O_2(G/L) \cap (H/L) = O_2(H/L) = 1.$$

So, $O_2(G/L) \subseteq C_{SL/L}(H/L) = 1$.

Lemma 4.5. *Suppose F is a proper subgroup of G that contains VT. Assume that $F \cap S$ is a Sylow 2-subgroup of F. Then $V \subseteq O_2(F)$.*

Proof. The proof is similar to that of Lemma 4.3.

Let $K = J_e(F \cap S)$. By Proposition 4.2(c),

$$(4.10) \qquad\qquad V \subseteq \Omega_1 Z(K).$$

Now suppose R is an arbitrary subgroup of $F \cap S$ that contains K and is an E-group. By (4.10), R contains V. Since $V \lhd S$, $V \lhd R$. By the choice of V and the definition of an E-group, $Z(R) \supseteq V$. Because of the arbitrary choice of R,

$$(4.11) \qquad V \subseteq \Omega_1(C_{F \cap S}(\hat{J}(F \cap S))) = \Omega_1 Z \hat{J}(F \cap S).$$

by Lemma 2.1.

Since $T \subseteq F$, we have $C_F(O_2(F)) \subseteq O_2(F)$. By induction,

$$F = C_F(\Omega_1 Z \hat{J}(F \cap S))N_F(K).$$

Therefore, by (4.10) and (4.11), $\langle V^F \rangle = \langle V^{N_F(K)} \rangle \subseteq \Omega_1 Z(K)$. Since $\langle V^F \rangle \lhd F$, $V \subseteq \langle V^F \rangle \subseteq O_2(F)$.

Now we come to the main goal of this section, the first major step in the proof of (D3).

Theorem 4.6. *We have* $V \subseteq T$.

Proof. Assume $V \not\subseteq T$. We will obtain a contradiction.

From Corollary 4.4, $|L/T|$ is odd. Therefore, $V \not\subseteq L$ and $VL/L \ne 1$. By Proposition 4.2 (d), Corollary 4.4 (h), and Lemma 4.5, and by Lemma 2.8 with SL/L, G/L, H/L, and VL/L in place of S, G, H, and V, there exists $x \in G$ such that

$$(4.12) \qquad G = \langle V, V^x, L \rangle.$$

Corollary 4.4 (f) asserts that $L/T \subseteq \Phi(G/T)$. By (4.12),

$$G/T = \langle VT/T, V^xT/T, \Phi(G/T) \rangle.$$

Therefore, Lemma I.2.5 yields that $G/T = \langle VT/T, V^xT/T \rangle$ and

$$(4.13) \qquad G = \langle V, V^x, T \rangle.$$

Since $T = \mathbf{O}_2(G)$ and $C(T) \subseteq T \subset G$, Lemma I.2.9 yields that there exists a non-central chief factor U/W of G within T. Choose $A \in \hat{\mathfrak{C}}_e(VT)$. Let $M = U/W$, $B = (A \cap T)^x$, and $q = |V/(V \cap T)|$. By Proposition 4.2 (c) and Lemma 2.1 (a), $V \subseteq \Omega_1 \mathbf{ZJ}_e(VT) \subseteq A$. Hence, by Dedekind's Lemma (Lemma I.5.2) $A = A \cap VT = V(A \cap T)$ and

$$(4.14) \qquad \begin{aligned} q &= |V/(V \cap (A \cap T))| = |V(A \cap T)/(A \cap T)| \\ &= |A/(A \cap T)| = |A|/|B|. \end{aligned}$$

Since $T = \mathbf{O}_2(G)$ and M is a chief factor of G within T, Lemma I.2.2 (f) yields that T centralizes M. As M is not a central chief factor of G, we see from (4.13) that $\langle V, V^x \rangle$ acts nontrivially on M. Since $V \triangleleft S$ and V is elementary Abelian,

$$(4.15) \qquad [U, V] \subseteq [S, V] \subseteq V$$

and V centralizes $[U, V]$ and $[M, V]$. It follows that $[M, V^x] = [M, V]^x \subseteq C_M(V^x)$ and, by (4.13),

$$(4.16) \qquad [M, V^x] \cap C_M(V) \subseteq C_M(V^x) \cap C_M(V) = C_M(G) = 1.$$

Now $B = (A \cap T)^x = A^x \cap T \supseteq V^x \cap T \supseteq V^x \cap U$. By (4.15), $[U, V^x] \subseteq V^x \cap U \subseteq B$. Therefore, (4.16) yields that

$$(4.17) \qquad |B/C_B(V)| \geq |[U, V^x]/([U, V^x] \cap C_U(V))| \geq |[M, V^x]|.$$

Let $r = |B/C_B(V)|$. From Corollary 4.4 (g) we have $V \cap T = V \cap C_S(M) = V \cap C_G(M)$. Hence $VC_G(M)/C_G(M) \cong V/(V \cap C_G(M)) = V/(V \cap T)$. Then by (4.17) and Theorem A,

$$(4.18) \qquad r \geq |[M, V^x]| = |[M, V]| \geq |VC_G(M)/C_G(M)|^2 = |V/(V \cap T)|^2 = q^2 > 1.$$

By the choice of V,

$$(4.19) \qquad |V/C_V(B)| > |B/C_B(V)|^{3/2} = r^{3/2}.$$

Since $C_B(V)V$ is an elementary Abelian group and $A \in \mathcal{C}_e(VT)$, (4.14) and (4.19) yield that

$$q|B| = |A| \geq |C_B(V)V| = |C_B(V)||V/(V \cap C_B(V))|$$

$$= r^{-1}|B||V/(V \cap B)| \geq r^{-1}|B||V/C_V(B)| > r^{1/2}|B|.$$

Therefore, $r < q^2$, contrary to (4.18). This contradiction completes the proof of Theorem 4.6.

5. Proof of (D3). In this section, we complete the proof of (D3). We retain the notation introduced at the beginning of §4. In particular, we assume that V satisfies the conditions of Proposition 4.1.

Since G is a counterexample to (D3), $G \neq N(J_e(S))$. Therefore, $J_e(S) \not\subseteq T$, by Lemma 2.1(b). Take $A \in \mathcal{C}_e(S)$ such that $A \not\subseteq T$. Let $Y = \langle V^G \rangle$ and $q = |A/(A \cap T)|$. By Theorem 4.6, $Y \subseteq T$.

One of the main ideas for dealing with $J_e(S)$ and similar characteristic subgroups is Thompson's idea of replacing one element of $\mathcal{C}_e(S)$ by another element which is 'better' in some sense. (Thompson's Replacement Theorem and a typical application are given as Theorems 14.2 and 14.3 of GL.) In §13 of his N-group paper [59], Thompson applies a similar technique to normal elementary Abelian 2-subgroups of certain local subgroups. Here we employ a variant of this method. We consider *all* the elementary Abelian subgroups of T and choose one such subgroup, B, for which $|B|^2|A \cap B|$ is maximal. This choice represents a compromise between choosing B to have maximal order and choosing B to have maximal intersection with A. In the proofs of the next two propositions, we 'replace' B by different subgroups and then use the maximality of B to complete the arguments.

Proposition 5.1. *We have* $Y \subseteq B$.

Proof. Assume this is false. Then there exists some $x \in G$ for which $V^x \not\subseteq B$. Let $y = x^{-1}$. By the maximal choice of B, the group $\langle B, V^x \rangle$ is not elementary Abelian. Therefore, B does not centralize V^x. Thus B^y does not centralize V. Let $r = |B/C_B(V^x)|$ and $B^* = V^x C_B(V^x)$.

By our choice of V,

$$|V^z/C_{V^x}(B)| = |V/C_V(B^y)| > |B^y/C_{B^y}(V)|^{3/2} = |B/C_B(V^x)|^{3/2} = r^{3/2}.$$

Consequently,

$$(5.1) \qquad |B^*| = |V^x C_B(V^x)| = |C_B(V^x)||V^x/(V^x \cap C_B(V^x))|$$

$$= r^{-1}|B||V^x/(V^x \cap B)| \geq r^{-1}|B||V^x/C_{V^x}(B)| > r^{1/2}|B|.$$

Since

$$r = |B/C_B(V^x)| \geq |(A \cap B)/(A \cap C_B(V^x))| \geq |A \cap B|/|A \cap B^*|,$$

$r|A \cap B^*| \geq |A \cap B|$. By (5.1), $|B^*| > r^{1/2}|B|$. Hence,

$$|B^*|^2|A \cap B^*| > r|B|^2|A \cap B^*| \geq |B|^2|A \cap B|,$$

contrary to the choice of B. This contradiction completes the proof of Proposition 5.1.

Lemma 5.2. *We have* $|Y/(Y \cap A)| \leq q^3$.

Proof. Since B is elementary Abelian and $A \in \mathcal{Q}_e(S)$,

(5.2) $$|A| \geq |B|.$$

By Proposition 5.1, B contains Y, whence

(5.3) $$|B/(A \cap B)| \geq |Y/(A \cap Y)|.$$

By the choice of B and by (5.2),

$$|B|^2|A \cap B| \geq |A \cap T|^2|A \cap T| = |A|^3q^{-3} \geq |B|^3q^{-3}.$$

Therefore, $q^3 \geq |B|/|A \cap B|$. By (5.3), we are done.

Lemma 5.3. *In each chief series of G that contains Y there is a unique noncentral chief factor of G within Y. Moreover, $A \cap T$ does not centralize Y.*

Proof. By Proposition 5.1, $Y \subseteq B$. Hence, Y is elementary Abelian. Since $A \in \mathcal{Q}_e(S)$, $C_Y(A) = A \cap Y$, by Lemma 2.1 (a). By Lemma 5.2,

(5.4) $$|Y/C_Y(A)| \leq q^3.$$

Take a composition series $Y = Y_0 \supset Y_1 \supset \cdots \supset Y_r = 1$ for Y under G. For $i = 1, \ldots, r$, let $F_i = Y_{i-1}/Y_i$. By (5.4) and Lemma 2.5,

(5.5) $$q^3 \geq |Y/C_Y(A)| \geq \prod_{1 \leq i \leq r} |F_i/C_{F_i}(A)|.$$

Suppose G centralizes every chief factor F_i. By Lemma I.2.9, $O^2(G)$ centralizes Y. By Proposition 4.2, $V \subseteq \Omega_1 Z(J)$. Therefore, by Corollary 4.4(a), $G = JO^2(G) \subseteq C(V)$. But then $V \subseteq Z(S)$, contrary to the choice of V.

Thus, there exists some chief factor F_j not centralized by G. Take such a factor F_j and let $C = C(F_j)$. By Corollary 4.4(g), $A \cap C = A \cap C_S(F_j) = A \cap T$ and $|AC/C| = |A/(A \cap C)| = |A/(A \cap T)| = q$. By Theorem A,

$$|F_j/C_{F_j}(A)| > |AC/C|^{3/2} \geq q^{3/2}.$$

By (5.5), F_j is the only chief factor within Y not centralized by G, and

(5.6) $$|Y/C_Y(A)| > q^{3/2}.$$

Let $D = C_{A \cap T}(Y)$. Since DY is elementary Abelian and $A \in \mathcal{Q}_e(S)$,

$$q|A \cap T| = |A| \geq |DY| = |D||Y/(Y \cap D)| \geq |D||Y/C_Y(A)|.$$

By (5.6), $q|A \cap T| > q^{3/2}|D|$ and $|A \cap T| > q^{1/2}|D| > |D|$. Therefore, $D \neq A \cap T$. This shows that $A \cap T$ does not centralize Y, which completes the proof of Lemma 5.3.

The next result completes the proof of Theorem B.

Lemma 5.4. *There is a contradiction.*

Proof. Let $U = \langle (A \cap T)^G \rangle$. By Proposition 4.2 and Lemma 2.1 (a),

$$(5.7) \qquad\qquad V \subseteq \Omega_1 Z(J) \subseteq A.$$

Consider a chief series of G that passes through Y and $C_Y(U)$. By Lemma 5.3, there exists a unique noncentral chief factor within Y in this series. This factor is involved within either $C_Y(U)$ or $Y/C_Y(U)$. Suppose it is involved within $C_Y(U)$. By Lemma I.2.9, $O^2(G)$ centralizes $Y/C_Y(U)$ and $G = O^2(G)S = C_G(Y/C_Y(U))S$. Since $V \lhd S$, $VC_Y(U) \lhd C_G(Y/C_Y(U))S = G$ and $Y = \langle V^G \rangle \subseteq VC_Y(U)$. By (5.7) and the definition of U, the group $A \cap T$ centralizes both V and $C_Y(U)$ and hence centralizes Y. But this contradicts Lemma 5.3. Consequently, the unique noncentral chief factor is involved within $Y/C_Y(U)$. By Lemma I.2.9,

(5.8) $O^2(G)$ centralizes $C_Y(U)$.

Since $\langle V^G \rangle = Y \supset C_Y(U)$,

$$(5.9) \qquad\qquad V \nsubseteq C_Y(U).$$

By (5.8) and Corollary 4.4 (a),

$$(5.10) \qquad\qquad G = O^2(G)J = C_G(C_Y(U))J.$$

By (5.7), $V \subseteq A \cap Y \subseteq A \cap T \subseteq U$. By (5.9), $V \nsubseteq Z(U)$. Then $1 \subset VZ(U)/Z(U) \lhd U/Z(U)$, from which it follows that

$$1 \subset (VZ(U)/Z(U)) \cap Z(U/Z(U)).$$

Take an element $x \in V - Z(U)$ for which the coset $xZ(U)$ lies in $Z(U/Z(U))$. Then

$$(5.11) \qquad\qquad 1 \subset [x, U] \subseteq V \cap Z(U) = V \cap C_Y(U).$$

Let $W = [x, U]$ and $Y_1 = \langle x^G \rangle$. Then $Y_1 \supseteq W$. By (5.7), (5.10), and (5.11),

$$(5.12) \qquad\qquad 1 \subset W \subseteq Z(G).$$

Since $x \in V - Z(U)$, (5.7) yields that $x \in Z(J)$ and (5.10) yields that $O^2(G)$ does not centralize x. Hence, $O^2(G)$ does not centralize Y_1. By Lemma I.2.9, Y_1 involves the unique noncentral chief factor within Y and $O^2(G)$ centralizes Y/Y_1. As $Y = \langle V^G \rangle = \langle (VY_1)^G \rangle$, (5.7) and (5.10) yield that

(5.13) $Y = VY_1$ and G centralizes Y/Y_1.

By (5.12) and the definition of W, $W \lhd G$ and $xW \in Z(U/W)$. Since $Z(U/W) \lhd G/W$, we have $Y_1/W \subseteq Z(U/W)$. Hence $A \cap T$ centralizes Y_1/W. Consequently, by (5.7) and (5.13), $A \cap T$ centralizes V and Y/W. Since $W \lhd G$ and $Y \lhd G$,

$$(5.14) \qquad C_G(Y/W) \supseteq \langle (A \cap T)^G \rangle = U.$$

By (5.14) and (5.12), $1 = [[Y, U], U] \supseteq [[V, U], U]$. Hence, $1 = [[V, U], U] = [[U, V], U]$. By the Three Subgroups Lemma, $1 = [[U, U], V] = [U', V]$. Since U is generated by conjugates of the elementary Abelian group $A \cap T$, it follows that $U/C_U(V)$ is elementary Abelian.

Let $\bar{U} = U/C_U(V)$. Then $\bar{U}$ operates faithfully on V. By (5.14) and the choice of V,

$$(5.15) \qquad |W| \geq |[Y, U]| \geq |[V, U]| = |[V, \bar{U}]| > |\bar{U}|.$$

By (5.12), the definition of W, and a short calculation,

$$[x, u_1 u_2] = [x, u_1][x, u_2] \quad \text{for all } u_1, u_2 \in U.$$

Therefore, the set $\{[x, u] | u \in \bar{U}\}$ is a group and thus is equal to W. It follows that $|W| \leq |\bar{U}|$, contrary to (5.15). This completes the proof of Lemma 5.4 and of Theorem B.

6. A strongly closed subgroup. In this section, we obtain a corollary to Theorem B.

Proposition 6.1. *Assume that $p = 2$ and G is S^4-free. Assume further that, whenever G^* is a section of G and $C_{G^*}(O_2(G^*)) \subseteq O_2(G^*)$, then every non-Abelian composition factor of G^* has the form $\mathrm{Sz}(2^{2n+1})$ or $\mathrm{PSL}(2, 3^{2n+1})$.*

Let $A = \langle (\Omega_1 Z(S))^{N(Je(S))} \rangle$.

Then A is an elementary Abelian strongly closed subgroup of S with respect to G and A controls strong fusion of S in G.

Proof. It is easy to see that we can define two section conjugacy functors W_1 and W_2 for p on G by $W_1(P) = J_e(P)$ and $W_2(P) = \Omega_1 Z \hat{J}(P)$ for all $P \in C_p^*(G)$. By Lemma 2.1, $\Omega_1 Z(S) \subseteq W_1(S) \cap W_2(S)$. Therefore, by Theorem B, W_1 and W_2 satisfy the hypothesis of Corollary I.5.7. The conclusion of Corollary I.5.7 yields the desired result.

7. Applications to simple groups. In the introduction to this chapter, we mentioned the following result:

(7.1) K_∞ controls strong fusion in G if $p \geq 3$ and every section of G is p-stable.

For $p = 2$, Proposition 6.1 yields an Abelian group A, nontrivial if $S \neq 1$, such that

(7.2) A controls strong fusion of S in G.

Unfortunately, this is only a weak analogue of (7.1) because the group A need not be a characteristic subgroup of S. Nevertheless, in some situations it actually yields more information than (7.1). The reason for this is a beautiful theorem of David Goldschmidt, given below, that applies only when $p = 2$. This theorem lists all simple groups in which a Sylow 2-subgroup possesses a nonidentity strongly closed Abelian subgroup.

Unfortunately, we will have to quote Goldschmidt's Theorem and some related results without proof. Up to this point, this monograph has been fairly self-contained except for results taken from GL. However, this section is an essential singularity.

In order to state Goldschmidt's Theorem, we require a definition. We say that G is a *group of Ree type* if G is simple and there exist an involution t of G, a subgroup K of G, and a natural number n such that $|G{:}C(t)|$ is odd $C(t) = \langle t \rangle \times K$, and $K \cong$ PSL$(2, 3^{2n+1})$. Here, it is easy to show that $C(t)$ contains an elementary Abelian Sylow 2-subgroup of G of order eight. As of this date, a great deal is known about groups of Ree type, but they are not completely determined.

By combining Goldschmidt's Theorem [34] with earlier papers of Janko [45], Janko and Thompson [46], J. Walter [62], and H. N. Ward [63] on groups with Abelian Sylow 2-subgroups, we obtain the following result.

Theorem 7.1. *Assume G is a non-Abelian simple group, $p = 2$, and there exists a nonidentity Abelian subgroup of S that is strongly closed in S with respect to G. Then either G is a group of Ree type or G is isomorphic to one of the following groups:*

    (a) $L_2(2^n), n \geq 3$;

    (b) $\text{Sz}(2^{2n+1}), n \geq 1$;

    (c) $U_3(2^n), n \geq 2$;

    (d) $L_2(q), q \equiv 3, 5 \pmod 8$ *and* $q \geq 5$;

    (e) J_1 *(Janko's first simple group).*

Here we define $L_2(q)$ and $U_3(q)$ for prime powers q as in [37, pp. 40 and 466]; thus, $L_2(q) = \text{PSL}(2, q)$. In [37], $U_3(q)$ is also denoted by $\text{PSU}(3, q)$. (Some authors denote the latter group by $\text{PSU}(3, q^2)$.) The group J_1 is of considerable historical importance because it was the first 'sporadic' simple group discovered since the Mathieu groups (announced in 1861 and 1873). It is defined and characterized in [45] (published in 1966).

We shall refer to the groups in Theorem 7.1 (including the groups of Ree type) as the *Goldschmidt groups*.

Lemma 7.2. *Suppose G is a Goldschmidt group. Then G is S^4-free. Moreover, the following conditions on G are equivalent:*

    (a) S^3 *is not involved in G;*

    (b) G *has the form* $\text{Sz}(2^{2n+1})$ *or* $\text{PSL}(2, 3^{2n+1})$.

Proof. By Lemma 2.6 (f), each of the groups in (b) satisfies (a) and hence is S^4-free. We must show that all the other Goldschmidt groups are S^4-free but violate (a). In fact, each of them contains a subgroup isomorphic to S^3.

Suppose $G \cong L_2(2^n)$ for some $n \geq 2$. By a theorem of Dickson [44, pp. 213–214], G has an elementary Abelian Sylow 2-subgroup and has dihedral subgroups of order $2z$ for every divisor z of $2^n \pm 1$. Since S^4 has a non-Abelian Sylow 2-subgroup, G

is S^4-free. Since 3 must divide $2^n + 1$ or $2^n - 1$, G has a dihedral subgroup of order 6.

Suppose $G \cong L_2(q)$ for some prime power q for which $q \geq 5$ and

(7.3) $q \equiv 3$ or 5 (modulo 8).

Then a Sylow 2-subgroup of G has order 4 and hence is Abelian. If q is not a power of 3, we may argue as in the previous case. If q is a power of 3, then, by (7.3), q is an odd power of 3; this case was handled at the beginning of the proof.

Suppose $G \cong U_3(2^n)$ for some $n \geq 2$. We apply Section II.10 of [44], where $U_3(2^n)$ is denoted by $PSU(3, 2^{2n})$. By considering an orthogonal basis $\{v_1, v_2, v_3\}$ of a unitary vector space and all permutations of the basis, we see that G contains a subgroup isomorphic to S^3. (Since $1 = -1$ in $GF(2^{2n})$, permutations yield linear transformations of determinant one.) If S^4 is involved in G, then by Lemma 2.3 there exists a nonidentity 2-subgroup T of G such that S^3 is involved in $N(T)$. Then we may assume that $T \subseteq S$. However, it follows from [44, p. 242], that S is a trivial intersection set in G. Therefore, $N(T) \subseteq N(S)$, which yields a contradiction because S^3 does not have a normal Sylow 2-subgroup.

Suppose $G \cong J_1$. By the definition of J_1 [45], G contains an involution whose centralizer is isomorphic to $Z_2 \times L_2(5)$. Hence G has a subgroup isomorphic to S^3. Since J_1 has an elementary Abelian subgroup of order 8, G is S^4-free.

Finally, assume G is of Ree type. As mentioned before Theorem 7.1, G has an elementary Abelian Sylow 2-subgroup and hence is S^4-free. By part (3) of the main theorem of [63, p. 63], some involution u of G normalizes but does not centralize some Sylow 3-subgroup T of G. By Lemma 2.2 applied to $\langle u, T \rangle$, the group G contains a subgroup isomorphic to S^3.

Now we may prove Theorem C. For the convenience of the reader, we repeat the statement of the theorem.

Theorem C. *Assume G is a non-Abelian simple group. The following are equivalent:*

(a) *G is S^4-free;*

(b) *G is a Goldschmidt group.*

Proof. By Lemma 7.2, every Goldschmidt group is S^4-free. We prove the converse by induction on $|G|$.

Assume G is S^4-free. By Proposition 6.1 and the definition of the Goldschmidt groups, it suffices to prove that, whenever G^* is a section of G for which

$$C_{G^*}(O_2(G^*)) \subseteq O_2(G^*),$$

then every non-Abelian composition factor of G^* has the form $Sz(2^{2n+1})$ or $PSL(2, 3^{2n+1})$. For such a section G^*, $|G^*| < |G|$, G^* is S^4-free, and $G^*/C_{G^*}(O_2(G^*)) = G^*/Z(O_2(G^*))$. Therefore, every composition factor of G^* is a Goldschmidt group by induction, and, by Lemma 2.3, S^3 is not involved in $G^*/Z(0_2(G^*))$. By Lemma 7.2, we are done.

Corollary 7.3. *Suppose G is a non-Abelian simple group. The following are equivalent:*

(a) *S^3 is not involved in G;*

(b) *G is isomorphic to $\mathrm{Sz}(2^{2n+1})$ or $\mathrm{PSL}(2, 3^{2n+1})$ for some natural number n.*

Proof. Use Theorem C and Lemma 7.2.

Corollary 7.3 includes Thompson's theorem [61] that the Suzuki groups are the only non-Abelian simple $3'$-groups. In addition, it permits us to generalize Theorem B as follows.

Theorem 7.4. *Suppose $p = 2$. Assume that*

(a) $C(0_2(G)) \subseteq 0_2(G)$, *and*

(b) *G is S^4-free.*

Then

$$G = C_G(Z(S))N(J_e(S)) = C(Z(S))N(\hat{J}(S)) = C(\Omega_1 Z \hat{J}(S))N(J_e(S)).$$

Proof. By (b) and Lemma 2.3, S^3 is not involved in $G/C(0_2(G))$. Therefore, by (a), S^3 is not involved in any composition factor of G. By Corollary 7.3, G satisfies condition (c) of Theorem B. Hence, Theorem B yields the desired factorizations.

Remark 7.1. Clearly, one may generalize Theorem A in a similar manner.

Now assume p is arbitrary. In Section 11 of GL, we discussed two groups which we denoted by $F(p)$ and $Qd(p)$. The group $F(p)$ is the set of all matrices of the form

$$\begin{bmatrix} a & b & c \\ 0 & a^{-1} & d \\ 0 & 0 & 1 \end{bmatrix} \quad \text{for which } a, b, c, d \in GF(p) \text{ and } a \neq 0.$$

The elements with $a = 1$ form a normal Sylow p-subgroup of $F(p)$ of order p^3 and index $p - 1$.

Generalizing the definition of $Qd(p)$, we define for each natural number n the 'quadratic' group $Qd(p^n)$ to be the subgroup of $\mathrm{SL}(3, p^n)$ consisting of all matrices of the form

$$(7.4) \qquad \begin{bmatrix} a & b & c \\ d & e & f \\ 0 & 0 & 1 \end{bmatrix} \quad \text{for which } a, b, c, d, e, f \in GF(p^n) \text{ and } ae - bd = 1.$$

It is easy to verify the following properties of $Qd(p^n)$.

Lemma 7.5. *Let n be a natural number. Then*

(a) *$0_p(Qd(p^n))$ is the set of all matrices of the form (7.4) with $a = e = 1$, $b = d = 0$, and c and f arbitrary;*

(b) *the matrices of the form (7.4) with $c = f = 0$ form a complement to $O_p(Qd(p^n))$ in $Qd(p^n)$ that is isomorphic to* $SL(2, p^n)$; *and*

(c) $Qd(p^n)$ *may be regarded as the semidirect product of a 2-dimensional vector space V over $GF(p^n)$ (i.e., $V = O_p(Qd(p^n)))$ by the group* $SL(V, GF(p^n))$.

Note that $Qd(2) \cong S^4$. Example 11.4 of GL shows that

(7.5) if $G = Qd(p)$, then $N(T) = N(S) \subset G$ for every nonidentity characteristic subgroup T of S.

Theorem 7.6. *Assume that p is odd. Then the following conditions on G are equivalent*:

(a) *The group $Qd(p)$ is not involved in G*;

(b) ZJ *controls strong fusion in every section of G*;

(c) K_∞ *controls strong fusion in every section of G*;

(d) *every section of G is p-stable*;

(e) *whenever G^* is a nonidentity section of G, S^* is a Sylow p-subgroup of G^*, and $C_{G^*}(O_p(G^*)) \subseteq O_p(G^*)$, then some nonidentity characteristic subgroup of S^* is a normal subgroup of G^*.*

Remark 7.2. The definition of ZJ is given on page 41 of GL.

Proof. Conditions (a), (b), and (c) are equivalent by Theorem 14.8 of GL; (a) and (d) are equivalent by Proposition 14.7 of GL. By (7.5), condition (e) yields condition (a). Finally, (d) yields (e) by (7.1) (Theorem 12.3(c) of GL).

Theorem 7.7. *Assume that $p = 2$. Then the following conditions on G are equivalent*:

(a) G *is S^4-free*;

(b) $\{Z, J_e\}$ *controls strong fusion in every section of G*;

(c) *whenever G^* is a nonidentity section of G, S^* is a Sylow p-subgroup of G^*, and $C_{G^*}(O_p(G^*)) \subseteq O_p(G^*)$, then*

$$G^* = N(A)N(B) = N(A)N(C) = N(B)N(C)$$

for some nonidentity characteristic subgroups A, B, C of S^*.

Proof. Assume (a). Then (b) follows from Theorem 7.4 and Corollary I.5.6, and (c) follows from Theorem 7.4.

Assume (a) is false, and let G^* be a section of G isomorphic to S^4. Then $G^* \cong Qd(2)$. By (7.5) and the definition of control of strong fusion, G^* is a counterexample to (b) and (c).

Theorem 7.6 and 7.7 yield a uniform result for all primes.

Theorem 7.8. *The following conditions on G are equivalent*:

(a) *The group $Qd(p)$ is not involved in G.*

(b) *There exists a set $\mathfrak{W}$ of section conjugacy functors such that, for every section G^* of G and every $W \in \mathfrak{W}$,*

(i) $\mathbb{W}$ *controls strong fusion in* G^*, *and*

(ii) *if* G^* *is a p-group, then* $\mathbb{W}(G^*)$ *is a characteristic subgroup of* G^*.

(c) *Whenever* G^* *is a section of* G, S^* *is a Sylow p-subgroup of* G^*, *and*
$C_{G^*}(O_p(G^*)) \subseteq O_p(G^*)$, *then*

$$G^* = \langle N(A) | A \text{ is a nonidentity characteristic subgroup of } S \rangle.$$

Proof. By Theorems 7.6 and 7.7, (b) and (c) follow from (a).

Assume (a) is false, and let G^* be a section of G isomorphic to $Qd(p)$. Let S^* be a Sylow p-subgroup of G^*. By (7.5), G^* violates (c). Since some element of $Z(S^*)$ is conjugate in G^* to an element of $O_p(G^*) - Z(S^*)$, we see by (7.5) that G^* violates (b).

Theorem 7.9. *Assume that* p *is odd and* $\mathbb{Z}J$ *does not control strong fusion in* G. *Let* $H = N(\mathbb{Z}J(S))$. *Then*

(a) $F(p)$ *is involved in* H, *and*

(b) $\mathbb{Z}J(S)$ *contains a cyclic subgroup* C *of order* p *such that* $|N_H(C)/C_H(C)| = p - 1$.

Proof. This is Theorem 14.13 of GL and is proved in [27].

Theorem 7.10 (Hayashi, Pettet). *The following conditions on* G *are equivalent:*

(a) *The group* G *has a normal Sylow 2-subgroup.*

(b) *For every nonidentity Sylow subgroup* T *of* G *of odd order,* $N(\mathbb{Z}J(T))$ *has a normal Sylow 2-subgroup.*

(c) *For every Sylow subgroup* T *of* G *of odd order, no nonidentity element of* T *is inverted by an element of* $N(\mathbb{Z}J(T))$.

Proof. It is easy to see (e.g., from the proof of Lemma 2.2) that (b) follows from (a) and that (c) follows from (b). To complete the proof, we will show that, if (a) is false, then so is (c).

Assume (a) is false. Then there exists $x \in G - O_2(G)$ such that $x^2 \in O_2(G)$. By Lemma I.2.10, there exists $g \in G$ such that $\langle x^g, x \rangle$ is not a 2-group. Then $\langle x^g, x \rangle O_2(G)/O_2(G)$ is a dihedral group and contains a dihedral subgroup of order $2q$ for some odd prime q. By Lemma 2.2, some nonidentity q-element y of G is inverted by some element of G. Let T be a Sylow q-subgroup of G that contains y. If $\mathbb{Z}J$ controls strong fusion for q in G, then y is inverted by an element of $N(\mathbb{Z}J(S))$. If $\mathbb{Z}J$ does not control strong fusion in G, then some nonidentity element of $\mathbb{Z}J(S)$ is inverted by an element of $N(\mathbb{Z}J(S))$, because of Theorem 7.9 and Lemma 2.2. Thus, in either case, (c) is false.

Theorem 7.10 and an application to fixed-point-free automorphisms were obtained independently by M. Hayashi [40] and M. Pettet [49]. Theorem 7.10 suggested the following result.

Theorem 7.11. *Suppose* T *is a Sylow 3-subgroup of* G. *Assume that no nonidentity element of* T *is inverted by an element of* $N(\mathbb{Z}J(T))$. *Then:*

(a) S^3 *is not involved in* G;

(b) ZJ *controls strong fusion in* G *for the prime* 3; *and*

(c) *every non-Abelian composition factor of* G *has the form* $Sz(2^{2n+1})$ *or* $PSL(2, 3^{2n+1})$.

Proof. The proof of Theorem 7.10 yields (a) and (b). Corollary 7.3 yields (c).

Remark 7.3. Suppose p is arbitrary and, whenever H is a p'-subgroup of G, then $N(H)/C(H)$ is a p'-group. The proof of Theorem 7.10 shows that $S \lhd G$ if $p = 2$. However, the situation can be much more complicated if p is odd. By using Theorem C, P. Ferguson [20] has determined the possible structure of G if $p = 3$. In particular, by Lemma 2.2 (with $p = 2$) and Theorem C, S^4 is not involved in G and every non-Abelian composition factor of G has the form $L_2(2^{2n+1})$ or $Sz(2^{2n+1})$.

Notes on Chapter II.

§1. The definition of $\hat{J}(T)$ is based on an idea suggested (independently) by S. Dolan and P. McBride.

§7. An interesting improvement of Theorem 7.6 has been obtained by L. Puig [52, p. 57]. He has shown that some conjugacy functor ZL controls strong fusion in G if p is odd and G (but not necessarily every section of G) is p-stable. He defines $ZL(P)$ to be $Z(L(P))$, where $L(P)$ is defined like $K^{\infty}(P)$ (in §13 of GL) with the following changes:

$\mathcal{B}(P)$ is replaced by the set of all Abelian subgroups of P;

$K^*(P; Q)$ and $K_*(P; Q)$ are both replaced by the set of all Abelian subgroups of P normalized by Q.

Chapter III. The General Situation

1. Introduction. At the end of the previous chapter, we summarized (Theorems 7.6–7.8) our main results on control of strong fusion. We may apply these results to the following problem:

(1.1) *Suppose G is contained as a maximal subgroup in a non-Abelian simple group H. Assume that $C_G(O_p(G)) \subseteq O_p(G)$ and $N_H(S) \not\subseteq G$. What can be said about G?*

Assume the conditions of (1.1). If p is odd, then Theorem II.7.6 yields that $Qd(p)$ is involved in G. If $p = 2$, then Theorem II.7.7 does not yield the same result except under additional restrictions, e.g., if $N_G(S)$ is contained in a unique maximal subgroup of G. In particular, if $p \doteq 2$ and $G/O_2(G)$ is a dihedral group of order $2q$ for some prime q, then $G/O_2(G) \cong S^3$. Unfortunately, our previous results tell us nothing further about G in this special case.

Given this analysis, it seems almost miraculous that there is a theorem of C. C. Sims [56] which applies to (1.1) precisely in the above special case. It asserts that G is isomorphic to S^4 or $S^4 \times Z_2$. Even after ten years of extensions and applications by many authors, this result is still not completely understood. In §2 we will describe a result of R. Niles which is the most general extension of Sims' result obtained so far. We conclude the chapter with some results on the global structure of G and some open questions. Unfortunately, space does not permit us to include any proofs.

2. Niles' Theorem. Recall that a *p-local* subgroup of G is the normalizer of a nonidentity p-subgroup; we call it a *maximal p-local subgroup* of G if it is contained in no other p-local subgroup of G. (Thus, G is a maximal p-local subgroup of itself if $O_p(G) \neq 1$.) With this notation consider the following problem.

(2.1) *Suppose G is contained as a p-local subgroup in a group H. Assume that $C(O_p(G)) \subseteq O_p(G)$ and $N_H(S) \not\subseteq G$. Can G be 'pushed up' to (shown to be contained in) some strictly larger p-local subgroup of H?*

In the situation of (1.1), the answer to the above 'pushing-up' problem is no. Consequently, as is easy to see,

(2.2) no nonidentity characteristic subgroup of S is normal in G.

Therefore, the observations in §1 show that $Qd(p)$ must be involved in G under certain conditions. By continuing with this approach, we may generalize (2.1) to a problem about G without reference to H. This has been done by Richard Niles [48] in a result which we will now discuss.

We first require some notation. Suppose L is a group isomorphic to $SL(2, p^n)$ for some natural number n. Assume that L operates faithfully on an elementary Abelian p-group V. We will say that V is a *standard module* for L if there exists a field F such that V is a 2-dimensional vector space over F and $SL(V, F)$ is the group of all automorphisms of V induced by L.

Theorem 2.1 (Niles). *Assume that G and S satisfy the following conditions:*

(a) $C(O_p(G)) \subseteq O_p(G)$:

(b) *for some normal subgroup K of G and some natural number n, $G/K \cong$* PSL$(2, p^n)$;

(c) *S is contained in a unique maximal subgroup of G; and*

(d) *no nonidentity characteristic subgroup of S is a normal subgroup of G.*

Then $G/O_p(G) \cong SL(2, p^n)$ and every noncentral chief factor of G within $O_p(G)$ is a standard module for $G/O_p(G)$. Moreover, G satisfies one of the following sets of conditions:

(I) (i) *S has nilpotence class 2;*

(ii) *S has exponent 4 if $p = 2$ and exponent p otherwise; and*

(iii) *G has only one noncentral chief factor within $O_p(G)$.*

(II) (i) *S has nilpotence class 3;*

(ii) *$p = 3$ and S has exponent 9; and*

(iii) *G has precisely two noncentral chief factors within $O_p(G)$.*

By using Lemma 7.5 and result (7.5) of Chapter II, one may easily show that $Qd(p)$ satisfies the hypothesis of Niles' Theorem. In fact, $Qd(p^n)$ likewise satisfies the hypothesis for every natural number n; the only difficult condition to check is (d). To verify it, let $H = SL(3, p^n)$ and $Q = Qd(p^n)$, and let U be the subgroup of H consisting of all upper triangular matrices (a_{ij}) for which $a_{11} = a_{22} = a_{33} = 1$. In addition, define τ to be the inverse-transpose automorphism of H and h to be the element

$$\begin{bmatrix} 0 & 0 & -1 \\ 0 & -1 & 0 \\ -1 & 0 & 0 \end{bmatrix}$$

of H. Let σ be the automorphism of H given by $x \to (x^\tau)^h$. Then U is a Sylow p-subgroup of H and of Q, and $U = U^\sigma$. A short calculation shows that $(O_p(Q))^\sigma \neq O_p(Q)$ and hence that no nonidentity characteristic subgroup of U is normal in Q, which proves (d).

By using the theory of groups of Lie type ([51, Chapter III]; [15]), we may modify and generalize the above proof of (d). Let Q^* be the group of all matrices (a_{ij}) in H for which $a_{31} = a_{32} = 0$. Then Q^* is a parabolic subgroup of rank 1 in H and $Q = O^{p'}(Q^*)$, $Q^* = QN_H(U)$, $U^\sigma = U$, and $(Q^*)^\sigma \neq Q^*$. Since H is associated with the Dynkin diagram A_2 of rank two, Q^* is a maximal subgroup of H. Therefore,

$$H = \langle Q^*, (Q^*)^\sigma \rangle = \langle Q, Q^\sigma, N_H(U) \rangle.$$

As $O_p(H) = 1$, we easily obtain (d).

With a few small changes, the above argument works just as well when A_2 is replaced by either of the other two Dynkin diagrams of rank two that admit graph automorphisms, namely, B_2 if $p = 2$ and G_2 if $p = 3$ ([51, p. 173]; [15, Chapter 12]). Both A_2 and B_2 yield examples of the first set of conditions in Theorem 2.1, while G_2 yields the second set. Actually, Niles' proof shows that *every* group which satisfies Theorem 2.1 can be built in a natural way from one of these three types of examples. In particular, he obtains the following application to the problem (1.1) of the previous section.

Theorem 2.2 (Niles). *Suppose G is a maximal p-local subgroup of a group H. Assume that G satisfies conditions* (a), (b), *and* (c) *of Theorem 2.1 and that* $N_H(S) \nsubseteq G$. *Then there exist subgroups G_0 and E of G such that E is an elementary Abelian p-group, $G = G_0 \times E$, and G_0 and p satisfy one of the following conditions:*

  (1) $G_0 \cong Qd(p^n)$;
  (2) $p = 2$ *and* $G_0 \cong O^{p'}(Q^*)$ *for every rank one parabolic subgroup of* $B_2(p^n)$;
  (3) $p = 3$ *and* $G_0 \cong O^{p'}(Q^*)$ *for every rank one parabolic subgroup of* $G_2(p^n)$.

Remark 2.1. Here, G need not be a maximal subgroup of H. Since the universal groups $B_2(2^n)$ $(= Sp(4, 2^n))$ and $G_2(3^n)$ have trivial centers, they are the same as the associated adjoint groups.

It is curious that, although the structure of G in Theorem 2.2 is described in terms of groups of Lie type of characteristic p, the group H in Theorem 2.2 cannot itself be a group of Lie type of characteristic p [13, Proposition 3.12]. In fact, frequently H is a group of Lie type for a different characteristic, e.g., when $p = 2$, $G \cong S^4$, and $H = PSL(2, 17)$.

Theorems 2.1 and 2.2 extend the theorem of Sims and some techniques of Dempwolff [16], W. Knapp [47], and the author ([26] and unpublished work). A very different proof of Theorem 2.1 for $p = 2$ was obtained independently by B. Baumann [10]. Both proofs use elementary methods but are slightly too long to be included here. (A small part of the proof of Theorem 2.1 was given at the Duluth Conference.) Both make essential use of the fact that G does not factor as $C(Z(S))N(J(S))$ (for $J(S)$ as defined in GL and [37]). By using some earlier techniques of Baumann, Niles has extended Theorem 2.1 to cover the situation in which G/K is not necessarily isomorphic to $PSL(2, p^n)$, but possesses a normal subgroup G^*/K such that $G^*/K \cong PSL(2, p^n)$ and $C_{G/K}(G^*/K) = 1$. In some work announced at the Duluth Conference [36], D. Goldschmidt has obtained an analogue to Theorem 2.1 with $p = 2$ and G/K of the form S^{2n+1} or A^{2n+1}. In addition, M. Aschbacher [5] has obtained some results related to Theorem 2.2.

Now suppose we move to the general case of Question 2, i.e., where we assume only that $C(O_p(G)) \subseteq O_p(G)$. As in Theorem 2.1, we wish to describe the situation in which G is badly behaved in some manner.

Assume first that $p \geq 5$ and that $K_\infty(S)$ or $ZJ(S)$ is not normal in G. As suggested by our discussion in Section II.1, there must be a chief factor M of G within $O_p(G)$

and a nonidentity element x of $G/C_G(M)$ for which $(x - 1)^2 = 0$, that is, $[[M, x], x] = 1$. Thompson's work on quadratic pairs ([60], [42]) completely determines the possible structure of the subgroup of $G/C_G(M)$ generated by all such elements x. In particular, it shows that this subgroup is built up from groups of Lie type acting on 'standard' or nearly 'standard' modules, such as occurs if $G = Qd(p^n)$ and $M = O_p(G)$ (Lemma II.7.5). However, almost all the groups of Lie type are generated by parabolic subgroups H which contain a given Sylow p-subgroup and satisfy the condition that $H/O_p(H) \cong$ SL$(2, p^n)$ for some n. Therefore, Theorem 2.1 yields partial information in this situation.

C.-Y. Ho has worked on quadratic pairs for $p = 3$. Although he has not completely determined them, his work suggests [43] that this case will be close to the preceding case.

Now suppose that $p = 2$. As usual, this case is unique. There are many well-behaved groups in which $K_\infty(S)$ or $ZJ(S)$ is not normal (GL, page 45; [25, Example 10.2]). Therefore, we might regard G as badly behaved if it violates the factorizations in Theorem B. In some very recent work [6], [7] M. Aschbacher has obtained analogues of the above results by assuming that

$$G \neq \langle C(\Omega_1 Z(S)), N(J_e(S)) \rangle,$$

or assuming the stronger condition that

(2.1) $G \neq \langle N(A) | A$ is a nonidentity characteristic subgroup of $S \rangle$.

(See Theorem II.7.8.) In place of the groups of Lie type, he obtains only groups of the form SL$(m, 2^n)$ and S^{2n+1}. His work uses Theorem 2.1 for $p = 2$ and many results about linear groups and simple groups.

3. **Global results.** In Chapter II, we classified the S^4-free simple groups. The bulk of the proof consisted of two separate results for $p = 2$. First, by making assumptions about groups involved in G, we showed (Proposition II.6.1) that S has a non-identity strongly closed Abelian subgroup. Then, given this strongly closed subgroup, Goldschmidt's Theorem provided the list of possible simple groups. These two examples illustrate the following two types of results:

(3.1) *Given some (apparently) weak local information about G, establish some strong local information about G.*

(3.2) *Given some strong local information about G, establish some global information about G.*

Many transfer results may also be viewed as combinations of these two types of results. For example, most of the proof of Theorem I.5.5 (b) consists of obtaining strong (local) information about the group

(3.3) $\langle xy^{-1} | x, y \in S$ and x and y are conjugate in $G \rangle$.

Then the Focal Subgroup Theorem and Proposition I.3.5 (used for Lemma I.5.3) yield global information about $S \cap G'$ which completes the proof.

As explained in the Preface, this book is primarily concerned with results of the first rather than the second type. In this section, however, we will quote some results of the second type. Since nearly all of the known results on simple groups are of the second type (e.g., classifications in which a Sylow 2-subgroup or the centralizer of an involution is given), we will restrict ourselves to a few results that are especially close to the material in this book.

Theorem II.7.1 is derived from a more general result:

Theorem 3.1 (Goldschmidt). *Suppose* $p = 2$ *and* A *is a strongly closed Abelian subgroup of* S *with respect to* G. *Let* $K = \langle A^G \rangle$ *and let* $\bar{H} = HO_{2'}(K)/O_{2'}(K)$ *for every subgroup* H *of* K. *Then*

  (a) $\bar{K} = O_2(\bar{K})\bar{K}'$ *and* $O_2(\bar{K}) \subseteq Z(\bar{K})$,

  (b) $\bar{A} = O_2(\bar{K})\Omega_1(\bar{T})$ *for some Sylow 2-subgroup* $\bar{T}$ *of* $\bar{K}$ *that contains* $\bar{A}$, *and*

  (c) $\bar{K}'/Z(\bar{K}')$ *is a direct product of non-Abelian simple Goldschmidt groups.*

Theorem 3.1 is based on an important sequence of papers by Suzuki, Bender, and Aschbacher. It extends an unpublished result of Shult (GL, page 62) and some earlier work of Goldschmidt [31], [33]. The following result of the author [24] was used in the proof of Theorem 3.1 and follows as a special case:

Theorem 3.2. *Suppose* $p = 2$, $x \in S$, *and* x *is weakly closed in* S *with respect to* G. *Define* $Z^*(G)$ *by*

$$Z^*(G)/O_{p'}(G) = Z(G/O_{p'}(G)).$$

Then $x \in Z^*(G)$.

For $p = 2$ and $Z^*(G)$ as in Theorem 3.2, the sequences $1 \subseteq O_{p'}(G) \subseteq Z^*(G)$ and $G \supseteq O^{p'}(G) \supseteq [O^{p'}(G), G]$ may be regarded as duals of each other. The Focal Subgroup Theorem shows that

$$O^{p'}(G)/[O^{p'}(G), G] \cong G/O^p(G)G' \cong S/(S \cap G')$$

and that $S \cap G'$ is given by (3.3). This suggests that, for $p = 2$, the weakly closed elements of S and the cosets in S of the group in (3.3) are somehow dual to each other, and Theorem 3.2 is dual to the Focal Subgroup Theorem (as is also illustrated in Appendix A1 of GL). It is not known whether Theorem 3.2 is valid for odd p, which is especially frustrating because here it is easy to test whether an element of S is weakly closed (GL, Theorems 12.7, 14.10).

Another application of strong closure occurs in the following results [35]:

Theorem 3.3 (Gorenstein-Harris, Goldschmidt). *Suppose* $p = 2$ *and* S *contains a subgroup of the form* $S_1 \times S_2$. *Assume that both* S_1 *and* S_2 *are strongly closed in* S *with respect to* G. *Then*

$$[\langle S_1^G \rangle, \langle S_2^G \rangle] \subseteq O_{2'}(G).$$

Corollary 3.4. *Suppose* $p = 2$ *and* S_1 *and* S_2 *are strongly closed subgroups of* S *with respect to* G. *Assume that* S_1 *and* S_2 *centralize each other. Then*

$$[\langle S_1^G \rangle, \langle S_2^G \rangle] \subseteq \langle (S_1 \cap S_2)^G \rangle O_{2'}(G).$$

In §1 we mentioned a theorem of Sims which determines G in a special case of (1.1) for $p = 2$; a theorem of W. J. Wong [67] actually determines the entire group H in this case. Other cases of (1.1) for $p = 2$ arise in the proof of the following result [9], which answers an open question in GL:

Theorem 3.5 (B. Baumann). *Suppose $p = 2$. Assume that S is a maximal subgroup of G and that $O_2(G) = O_{2'}(G) = 1$. Then $O^{2'}(G)$ is a direct product of simple groups with dihedral Sylow 2-subgroups.*

G. Higman and his school have developed the subject of "odd characterizations" [41], i.e., results of type (3.2) for odd p. By using global results and methods for $p = 2$ and special arguments for $p = 3$, they have succeeded in proving analogues for $p = 3$ of some of the early results for $p = 2$. A recent result in this area, which uses the classification of S^4-free groups, is the proof [22] by L. R. Fletcher, B. Stellmacher, and W. B. Stewart that every non-Abelian simple group with no elements of order six has the form $L_2(q)$, $Sz(q)$, $L_3(2^n)$, or $U_3(2^n)$. This includes the case of simple groups in which the centralizer of every element of order 3 is a 3-group.

4. Open questions. The results in this book leave a number of open questions. The outstanding one seems to be the same as in GL, namely, to find a substitute for K_∞ and ZJ in Theorem II.7.6.

Question 4.1. Suppose that P is a nonidentity finite 2-group. Does there exist a nonidentity characteristic subgroup $L(P)$ of P such that $L(P) \triangleleft H$ for every S^4-free group H that contains P as a Sylow 2-subgroup and satisfies the restriction that $C_H(O_2(H)) \subseteq O_2(H)$?

As mentioned in GL (p. 49), many special cases of this question are also of interest. Unfortunately, the candidate for $L(P)$ mentioned in GL does not work; S. Dolan has pointed out a counterexample in which $O_2(H)$ is extra-special of order 2^5 and $H/O_2(H)$ is dihedral of order 10.

The surprising new results of Yoshida suggest that it may well be possible to determine more information about transfer in general (see also [53], [54]). While the case in which $p = 2$ seems the most difficult and important, the author does not know of any specific conjectures. For $p = 3$, we may generalize a question in GL as follows:

Question 4.2. Suppose that $p = 3$. Does there exist a set of conjugacy functors that controls transfer in G?

Of course, we would prefer a set $\mathfrak{W}$ such that $\{W(S) | W \in \mathfrak{W}\}$ is a set of characteristic subgroups of S which can be chosen independently of G. We may similarly generalize two related open questions (16.4 and 16.8) in GL. For Question 16.4 we assume that $\mathfrak{W}$ does not control strong fusion in G; for Question 16.8, we ask that a cohomology class of S be stable [14, pp. 256–259] in G if it is stable in $N(W(S))$ for every $W \in \mathfrak{W}$. (*Added in proof.* The author has answered Question 4.2 affirmatively in unpublished work.)

The material in §§1 and 2 yields some special cases of the following conjecture:

Question 4.3 (C. C. Sims). Does there exist a function f on the natural numbers enjoying the following property: Whenever M is a maximal subgroup of G, $x \in G - M$, $n = |M: M \cap M^x|$, and M contains no nonidentity normal subgroup of G, then $|M| \leq f(n)$?

Thompson and Wielandt have shown [65, Theorem 6.7], that, for some prime p depending on M, x, and G,

$$|M/O_p(M)| \leq (n + 1)!(n - 1)!^n.$$

All of the above questions might be simplified or solved if the next question could be answered:

Question 4.4. Assume conditions (a), (b), and (c) of Theorem 2.1. Describe the largest characteristic subgroup of S which is normal in G.

Note that, modulo this subgroup, G satisfies the conclusion of Theorem 2.1. Hence, the subgroup contains $(S')'$, and much more can be said about it.

The discussion of Theorem 3.2 raises the following question:

Question 4.5. Is Theorem 3.2 valid if p is odd?

This question appeared in GL, and the special case of Shult [55] mentioned there is still the only case known. We may amplify another question in GL as follows:

Question 4.6. Assume p is odd and $N(S)/C(S)$ is a p-group. Do $\mathbf{ZJ}$, $\mathbf{K}_\infty$, or $\mathbf{K}^\infty$ control strong fusion in G? Is G p-solvable if $p > 3$? Can G be simple if $p = 3$ and $S \neq 1$?

We note that G need not be p-solvable if $p = 3$, e.g., if G is the semidirect product of $PSL(3, 3^3)$ by its group of field automorphisms.

Questions 16.2 (a) and 16.2 (b) of GL are answered in the affirmative and in the negative respectively in Theorem III.3.5 of this work and in [28]. As mentioned in page 63 of GL and §III.2 of this work, Question 16.5 of GL has been answered completely for $p \geq 5$ and partially for $p = 3$.

In the previous section, we described two general types of results, local and global ((3.1) and (3.2)), and quoted some of the global results. A comparison of this section with the local results in the rest of the book suggests some general problems. Most local results have been proved for all odd p (or at least all $p \geq 5$) and are false or much more difficult for $p = 2$ (e.g., see §§11 and 12 of GL and compare the proof of Theorem 12.9 of GL with that of Theorem B). Thus we know least about p-local subgroups and fusion for $p = 2$. On the other hand, most global results have been proved only for $p = 2$, and in recent years the proofs have increasingly depended upon 2-local subgroups and fusion in the Sylow 2-subgroup. (By Aschbacher's work [4], one may frequently assume that $C_L(O_2(L)) \subseteq O_2(L)$ for every 2-local subgroup L, thus bypassing many problems about the core, $O_{2'}(L)$, in earlier papers.) Indeed, the 2-local structure is known [1] to determine the order of G in general unless $|G|$ is odd or G has only one conjugacy class of involutions.

It is hard to argue with a counterexample. Therefore, we must accept the fact that fusion for Sylow 2-subgroups is intrinsically more complicated than for other Sylow subgroups and strive for whatever advances are possible. For example, Yoshida's transfer results, although formally weakest for $p = 2$, have already had valuable applications in this case [68], [69].

On the other hand, the situation for global results is quite different. Here, as the pages of proof mount into the thousands, many of them using similar laborious methods for similar problems, one suspects that there must be additional methods, not yet discovered.

(Added in Second Printing)

There has been significant progress on the open questions stated above and in GL. Here we describe the answers obtained since the first printing.

The most striking advance has been an affirmative answer to Question 4.1 by Bernd Stellmacher, which we describe in Appendix A3.

We regret that the answer to Question 4.2 mentioned in the first printing has remained unpublished. The author hopes to remedy this situation soon.

Question 4.3 has been answered affirmatively in [70], but only by assuming the classification of finite simple groups (which we denote by CFSG).

Let S be a p-group. As a first step toward question 4.4, let K be the largest characteristic subgroup of S that is normal in G for *every* group G as in Question 4.4. A family of examples in pp. 412–413 of [72] shows that one may have $K = 1$ even if S has arbitrarily large nilpotence class. (In fact, every finite p-group is isomorphic to a subgroup of one of the groups in this family.) Therefore, there is no general solution to Question 4.4. However, by using Theorem B of [72] and taking iterated factor groups, one can construct (in principle) a specific family of characteristic subgroups L of S such that

(i) for each L, S/L is "small" in the sense that it satisfies parts (i) and (ii) of Theorem 2.1(I) or of Theorem 2.1(II), and

(ii) for each G, one of the groups L is normal in G.

Question 4.5 has been answered affirmatively in [74], Remark 7.8.3, but only by assumiing CFSG.

Question 16.4 of GL asks whether there exists a section conjugacy functor on G of degree at most 2 if $p \geq 3$, or degree at most 3 if $p \geq 5$. An affirmative answer to the second part is given in Theorem 3 of [71].

For Question 16.5 the cases when $|V/C_V(A)| \leq |A|$ for $p = 3$ and A acting quadratically can apparently be determined by assuming CFSG and combining unpublished work of A. Chermak or U. Meierfrankenfeld with previous work of Premet and Suprunenko [77]. As of this writing, it appears that the analogous cases for $p = 2$ can be determined similarly by published and unpublished work of a variety of authors.

Question 16.8 (on control of cohomology groups $H^n(G, A)$ for a trivial G-module A of order p) has been solved for $n = 2$ by D. Holt [75] and for general n by M. Miyamoto [76]. A slight improvement in their bounds is given in page 305 of [71].

Appendix A1. Proof of Theorem A

In this section, we prove Theorem A of Chapter II. The proof requires a number of preliminary results beyond those of Section II.2.

Lemma 1.1 (‘$P \times Q$’ Lemma; **J. G. Thompson**). *Suppose M is a p-group and G is a group of automorphisms of M. Assume that $G = P \times Q$ for some p-group P and some group Q for which $O_p(Q) = 1$. Then Q acts faithfully on $C_M(P)$.*

Proof. Suppose x is a p'-element of Q that centralizes $C_M(P)$. Then $P \times \langle x \rangle$ acts faithfully on M and $\langle x \rangle$ acts trivially on $C_M(P)$. By Theorem 5.3.4 of [37], $\langle x \rangle = 1$.

Since x was chosen arbitrarily above, $C_Q(C_M(P))$ is a p-group. As $C_Q(C_M(P))$ $\lhd \, Q$ and $O_p(Q) = 1$, $C_Q(C_M(P)) = 1$. Thus, Q acts faithfully on $C_M(P)$.

Lemma 1.2. *Suppose p is an odd prime, M is an elementary Abelian 2-group, G is a subgroup of Aut M, u is an involution in G, and H is a nonidentity elementary Abelian p-subgroup of G inverted by u. Let e be the smallest natural number for which p divides $2^{2e} - 1$. Let $p^n = |H|$. Then*
   (a) $|M/C_M(u)| \geq 2^{en}$;
   (b) *if* $p \geq 5$, *then* $|M/C_M(u)| \geq 2^{2n}$;
   (c) *if* $p > 5$, *then* $|M/C_M(u)| \geq 2^{3n}$; *and*
   (d) *if* $C_M(H) = 1$, *then* $|M| = |C_M(u)|^2$.

Proof. We use induction on $|M|$.

Easy calculation shows that (b) and (c) follow from (a). Therefore, we need to prove only (a) and (d). Clearly, we may assume that $G = \langle H, u \rangle = H\langle u \rangle$.

Now, G normalizes $[M, H]$. By Lemma I.2.7, $M = [M, H] \times C_M(H)$. Therefore, H, and hence G, act faithfully on $[M, H]$. Since

$$|M/C_M(u)| \geq |[M, H]/C_{[M,H]}(u)|,$$

we may assume that

(1.1) $$M = [M, H] \quad \text{and} \quad C_M(H) = 1.$$

Let M_0 be a minimal H-invariant subgroup of M, and let $K = C_H(M_0)$. Then H/K acts faithfully and irreducibly on M_0. Therefore, H/K is cyclic. By (1.1),

(1.2) $$|H/K| = p.$$

By Lemma I.2.7,

(1.3) $$M = [M, K] \times C_M(K).$$

Since u inverts K, it follows that $K \triangleleft G$ and that G fixes $[M, K]$ and $C_M(K)$.

Assume $K \neq 1$. Then $[M, K] \neq 1$. Since $1 \subset M_0 \subseteq C_M(K)$,

(1.4) $$|[M, K]| < |M| \quad \text{and} \quad |C_M(K)| < |M|.$$

Let $L = C_H([M, K])$ and $p^r = |H/L|$. Since $K \subseteq C_H(C_M(K)) \subseteq C_H(M_0) = K$, $K = C_H(C_M(K))$. Hence, by (1.3),

$$L \cap K = C_H([M, K]) \cap C_H(C_M(K)) = C_H(M) = 1.$$

Therefore, $p^n = |H| = |H/(L \cap K)| \leq |H/L||H/K| = p^r \cdot p = p^{r+1}$, and $n \leq r + 1$.

Continuing in the above situation, we note that G/L and G/K act faithfully on $[M, K]$ and $C_M(K)$ respectively. By (1.1), (1.4) and induction,

$$|[M, K]/C_{[M, K]}(u)| \geq 2^{er}, \quad |C_M(K)/(C_M(K) \cap C_M(u))| \geq 2^e,$$

$|[M, K]| = |C_{[M, K]}(u)|^2$, and

$$|C_M(K)| = |C_M(K) \cap C_M(u)|^2.$$

By (1.3), we obtain (d) and also obtain that

$$|M/C_M(u)| \geq 2^{er} \cdot 2^e = 2^{e(r+1)} \geq 2^{en},$$

as desired.

Finally, we assume that $K = 1$. By (1.2), $|H| = p$ and G is a dihedral group. Take $y \in H^\#$. For each $x \in M$, we have $[x, u] = xx^u \in C_M(u)$. Since the mapping ϕ given by $x \to [x, u]$ is an endomorphism of M, $|M/C_M(u)| = |M/\text{Ker } \phi| = |\text{Im } \phi| \leq |C_M(u)|$. Thus, $|M| \leq |C_M(u)|^2$. Since $G = \langle u, u^y \rangle$, equation (1.1) yields that

$$|M| = |M/C_M(\langle u, u^y \rangle)| \leq |M/C_M(u)||M/C_M(u^y)| = |M|^2/|C_M(u)|^2.$$

Therefore, $|C_M(u)|^2 \leq |M|$, from which (d) follows.

Let $|C_M(u)| = 2^f$. By (d), $|M| = 2^{2f}$. Since $C_M(H) = 1$ by (1.1),

$$2^{2f} \equiv |M| \equiv 1 \pmod{p}.$$

Therefore, $f \geq e$ and $|M/C_M(u)| = 2^f \geq 2^e$, which proves (a). This completes the proof of Lemma 1.2.

Lemma 1.3 (Zsigmondy). *Suppose a and k are natural numbers greater than one. Assume that*

(a) *if $k = 6$, then $a > 2$, and*

(b) *if $k = 2$, then $a + 1$ is not a power of 2.*

Then there exists a prime q such that $q|(a^k - 1)$ and for $i = 1, \ldots, k - 1$, $q \nmid (a^i - 1)$.

Proof. This is a well-known result which follows from Corollary 2, page 358, of [3].

Lemma 1.4. *Suppose M is an elementary Abelian p-group, G is a subgroup of Aut M, and $H \lhd G$ and $L = C_M(C_S(H))$. Assume that $O_p(H) = 1$. Then $O_p(HS/C_S(H)) = 1$ and $HS/C_S(H)$ acts faithfully on L.*

Proof. Since

$$H \cap C_S(H) \subseteq O_p(Z(H)) \subseteq O_p(H) = 1,$$

$HC_S(H) = H \times C_S(H)$. By Lemma 1.1,

(1.5) H acts faithfully on L.

Now, HS is a subgroup of G in which S is a Sylow p-subgroup. Since $C_S(H) \lhd HS$, it follows that HS acts on L. By (1.5),

$$C_{HS}(L) \cong C_{HS}(L)H/H \subseteq HS/H$$

and $C_{HS}(L) \subseteq O_p(HS) \subseteq S$. Take $S_1 \subseteq S$ such that $S_1 \supseteq C_{HS}(L)$ and $S_1/C_{HS}(L) = O_p(HS/C_{HS}(L))$. Then $C_S(H) \subseteq C_{HS}(L) \subseteq S_1$, $S_1 \lhd HS$, and $[H, S_1] \lhd H$. Therefore, $[H, S_1] \subseteq O_p(H) = 1$ and $S_1 \subseteq C_S(H)$. Thus, $C_S(H) = C_{HS}(L) = S_1$, which completes the proof of the lemma.

Lemma 1.5. *Suppose G is a group and n is a natural number. Assume that $O_2(G) = 1$ and that $G/Z(G)$ is a direct product of simple groups, either all isomorphic to $Sz(2^{2n+1})$ or all isomorphic to $PSL(2, 3^{2n+1})$. Then*

$$G = G' \times Z(G).$$

Proof. We use induction on $|G|$. Note that

(1.6) G has no composition factors of order 2.

Assume first that $G/Z(G)$ is not simple or that $G' \subset G$. Then G has a proper non-Abelian normal subgroup N. Let H be the subgroup of G for which $H \supseteq N'$ and $H/N' = (G/N)'$. By (1.6), $O_2(G/N') = 1$ and $O_2(N) = 1$. Since $Z(G) \cap N \subseteq Z(N)$ and

$$N/(Z(G) \cap N) \cong NZ(G)/Z(G) \lhd G/Z(G),$$

the induction hypothesis applies to N. Hence, $N = N' \times Z(N)$. Similarly,

$$G/N' = (H/N') \times Z(G/N').$$

Therefore, H has no cyclic composition factors and $G = HZ(G)$. Hence, $H = H' = G'$ and $G = H \times Z(G)$, as desired.

Thus, we may assume that $G/Z(G)$ is simple and $G' = G$. We will assume that $Z(G) \neq 1$ and obtain a contradiction. Take a subgroup Y of some prime index in $Z(G)$. Assume the index is p. Then p is odd and $G/Y = G'/Y = (G/Y)'$. By induction, $Y = 1$. Hence,

(1.7) $$|Z(G)| = p \geq 3.$$

Since S is a Sylow p-subgroup of G, $S = S \cap G = S \cap G'$. If S is Abelian, then by Theorem I.3.6 and Lemma I.2.7,

$$S = S \cap G' = [S, N(S)] \quad \text{and} \quad S = (S \cap G') \times (S \cap Z(N(S))).$$

But then $1 = S \cap Z(N(S)) \supseteq Z(G) \supset 1$, a contradiction. Therefore,

(1.8) S is not Abelian.

Since S is not Abelian, $S/Z(S)$ is not cyclic. Consequently, $S/Z(G)$ is not cyclic. From Lemma II.2.6 and the structure of $PSL(2, 3^{2n+1})$, it follows that

$$G/Z(G) \cong PSL(2, 3^{2n+1})), \quad p = 3,$$

and $S/Z(G)$ is an elementary Abelian group of order 3^{2n+1} on which $N_G(S)$ acts irreducibly by conjugation. By (1.7) and (1.8), $S' = \Phi(S) = Z(S) = Z(G)$. Thus, S is an extra-special group, which is impossible because $|S/Z(S)|$ is an odd power of 3 [37, pp. 183 and 204].

Lemma 1.6. *Suppose G is an operator group on an elementary Abelian p-group M. Let L be the dual module for M over Z_p, i.e.,*

$$L = \text{Hom}(M, Z_p) \quad \text{and} \quad f^g(x) = f(x^{g^{-1}}) \quad \text{for all } f \in L, x \in M, g \in G.$$

Then $|L/C_L(G)| = |[M, G]|$.

Proof. We regard M and L as vector spaces over Z_p and hence as additive groups. By the elementary properties of vector spaces, it suffices to show that $C_L(G)$ is the annihilator of $[M, G]$ in L, in which case the dimensions of $L/C_L(G)$ and $[M, G]$ over Z_p are equal.

Let $x \in M$, $f \in L$, and $g \in G$. If $f \in C_L(G)$, then

$$f(x^{-1}x^g) = -f(x) + f(x^g) = f^{g^{-1}}(x) - f(x) = f(x) - f(x) = 0.$$

On the other hand, if f annihilates $[M, G]$, then

$$f^g(x) - f(x) = f(x^{g^{-1}}) - f(x) = f(x^{-1}x^{g^{-1}}) = 0.$$

Now we begin the proof of Theorem A. For the sake of convenience, we repeat its statement.

Theorem A. *Suppose M is an elementary Abelian 2-group and G is a subgroup of Aut M. Assume that*

(a) $O_2(G) = 1$,

(b) S^3 *is not involved in G, and*

(c) *every non-Abelian composition factor of G has the form*

$$Sz(2^{2n+1}) \quad or \quad PSL(2, 3^{2n+1})$$

Then, for every elementary Abelian 2-subgroup A of G,

$$|M/C_M(A)| \geq |A|^2 \quad and \quad |[M, A]| \geq |A|^2.$$

Proof of Theorem A. We use induction on $|G|$. The result is obvious for $A = 1$; therefore, we take A to be a nonidentity elementary Abelian 2-subgroup of G. By

Lemma 1.6, it suffices to prove that $|M/C_M(A)| \geq |A|^2$ because the dual of M also satisfies the hypothesis on M.

Obviously, condition (b) is inherited by every subgroup of G. It is an elementary fact that, for any subgroup of any group, the composition factors of the subgroup are isomorphic to sections of the composition factors of the group. Therefore, by Lemma II.2.6, condition (c) is inherited by every subgroup of G.

Since $0_2(Z(G)) \subseteq 0_2(G) = 1$, $A \not\subseteq Z(G)$. Take $H \lhd G$ minimal such that A does not centralize H.

Let S be a Sylow 2-subgroup of G that contains A and let $L = C_M(C_S(H))$. By hypothesis,

$$(1.9) \qquad\qquad 0_2(H) \subseteq 0_2(G) = 1.$$

By (1.9) and Lemma 1.4,

(1.10) $HS/C_S(H)$ acts faithfully on L and $0_2(HS/C_S(H)) = 1$.

Suppose $HS \subset G$. Since $H \not\subseteq Z(G)$, $C(H)$ is a proper normal subgroup of G. Then $0_2(C(H)) \subseteq 0_2(G) = 1$ and, by induction,

$$|M/C_M(C_A(H))| \geq |C_A(H)|^2.$$

Then, by (1.10) and induction,

$$
\begin{aligned}
|M/C_M(A)| &= |M/C_M(C_A(H))| |C_M(C_A(H))/C_M(A)| \\
&\geq |C_A(H)|^2 |C_L(C_A(H))/C_L(A)| \\
&= |C_A(H)|^2 |L/C_L(AC_S(H)/C_S(H))| \\
&\geq |C_A(H)|^2 |AC_S(H)/C_S(H)|^2 \\
&= |C_A(H)|^2 |A/C_A(H)|^2 = |A|^2.
\end{aligned}
$$

As mentioned above, this inequality suffices to yield the conclusion.

For the remainder of the proof, we will assume that $HS = G$. Then G/H is a 2-group. Consequently, HA is a subnormal subgroup of G and $0_2(HA) \subseteq 0_2(G) = 1$. If $HA \subset G$, then the conclusion follows by induction. Therefore, we assume that

$$(1.11) \qquad\qquad HA = G.$$

Now, $[H, A] \lhd HA = G$. Suppose $[H, A] \subset H$. Then A centralizes $[H, A]$ by our choice of H, and $A[H, A] \lhd HA = G$, $1 \subset A \subseteq 0_2(Z(A[H, A])) \subseteq 0_2(G) = 1$, a contradiction. Therefore,

$$(1.12) \qquad\qquad [H, A] = H.$$

Take $K \lhd G$ maximal such that $K \subset H$. Then A centralizes K and $C_G(K) \lhd G$. By (1.11) and (1.12),

$$G = HA = [H, A]A \subseteq \langle A^H \rangle \subseteq \langle (C_G(K))^H \rangle = C_G(K).$$

Hence,

(1.13) $$K \subseteq Z(G) \cap H \subseteq Z(H).$$

By our choice of K,

(1.14) H/K is a chief factor of G.

Suppose H/K is Abelian. Then $H' \subseteq K \subseteq Z(H)$, so H is nilpotent. By the choice of H, there exists a prime q for which H is a q-group. By (1.13), $C_A(H/K)$ stabilizes the normal series $H \supseteq K \supseteq 1$. Therefore, $C_A(H/K) \lhd HA = G$, $C_A(H/K) \subseteq O_2(G)$ $= 1$, and A acts faithfully on H/K. By (1.11) and (1.14), A acts irreducibly on H/K. Hence, $|A| = 2$ and $|H/K| = q$. By (1.13), $H/Z(H)$ is cyclic, and so H is Abelian. By Lemma I.2.7 and the choice of H, $H = [H, A] \times K$, $H = [H, A]$, and $K = 1$. Thus, $|H| = q$ and G is dihedral of order $2q$. By Lemma 1.2(b),

$$|M/C_M(A)| \geq 2^2 = |A|^2,$$

as desired.

Henceforth, we assume that H/K is not Abelian. By (1.14), H/K is a direct product of isomorphic simple groups. By hypothesis, these groups have the form $Sz(2^{2n+1})$ or $PSL(2, 3^{2n+1})$. By (1.13) and Lemma 1.5,

$$H = H' \times Z(H) = H' \times K.$$

Now, $H' \lhd G$. By (1.12) and (1.13), $H = [H, A] = [H', A] \times [K, A] = [H', A] \subseteq H'$. Therefore, $H' = H$ and $K = 1$. Consequently,

(1.15) H is a minimal normal subgroup of G

and

(1.16) H is a direct product of isomorphic groups of the form $Sz(2^{2n+1})$ or $PSL(2, 3^{2n+1})$.

By (1.11), G/H is a 2-group. Therefore,

(1.17) $$H = O^2(G).$$

We now divide the proof into two cases, according to the two possibilities in (1.16).

Case 1. The simple direct factors of H have the form $Sz(2^{2n+1})$. Here we let $q = 2^{2n+1}$. Let $J_1, \ldots, J_r$ be subgroups of H isomorphic to $Sz(q)$ such that $H = J_1 \times \cdots \times J_r$. By (1.17), $H = O^2(G)$. By hypothesis, $O_2(G) = 1$. Therefore, by Lemma II.2.7,

(1.18) $$\bigcap_{1 \leq i \leq r} N_G(J_i) = H.$$

By (1.15) and (1.11), A permutes the groups $J_1, \ldots, J_r$ transitively by conjugation. By (1.18), AH/H acts faithfully on the set $\{J_1, \ldots, J_r\}$. As AH/H is Abelian, AH/H acts as a regular permutation group on this set. Therefore, $|AH/H| = r$.

Let T be a transversal to $A \cap H$ in A. Let

$$J = \left\{ \prod_{t \in T} x^t \,\middle|\, x \in J_1 \right\}.$$

Then $J \cong J_1 \cong Sz(q)$ and a short argument shows that $A \cap H \subseteq C_H(A) \subseteq J$.

Let B be a maximal cyclic subgroup of $A \cap H$. Since A is elementary Abelian, $|B| \leq 2$. Recall that $q = 2^{2n+1}$. By Zsigmondy's result (Lemma 1.3) there exists a prime s such that

(1.19) $s|(2^{4(2n+1)} - 1)$ and, for $i = 1, 2, \ldots, 4(2n + 1) - 1$, $s \nmid (2^i - 1)$.

Then $s|(q^2 + 1)$ because $s|(q^4 - 1)$ and $s \nmid (q^2 - 1)$. By Lemma II.2.6(e), J contains a cyclic subgroup R of order s such that B inverts R if $B \neq 1$.

For $i = 1, \ldots, r$, let R_i be the image of R under the natural projection of H onto J_i. Let $R^* = R_1 \times \cdots \times R_r$. It is easy to see that

(1.20) B inverts R^* if $B \neq 1$.

Suppose $B = 1$. Then $A \cap H = 1$ and A acts as a regular permutation group on the set $\{R_1, \ldots, R_r\}$ by conjugation. Therefore, R^*A is a proper subgroup of G and $O_2(R^*A) = 1$. By induction,

$$|M/C_M(A)| \geq |A|^2,$$

as desired.

Now suppose $A \cap H \neq 1$. Then $|B| = 2$. Since $A \cap H \subseteq J$ and $J \cong Sz(q)$, we have $|A \cap H| \leq q$. Hence,

$$|A| = |A/(A \cap H)||A \cap H| = r|A \cap H| \leq rq.$$

By (1.19), (1.20), and Lemma 1.2(a), $|M/C_M(B)| \geq q^{2r}$. Therefore,

$$|M/C_M(A)| \geq |M/C_M(B)| \geq q^{2r}$$

and

$$\frac{|M/C_M(A)|}{|A|^2} \geq \frac{q^{2r}}{|A|^2} \geq \frac{q^{2r}}{r^2|A \cap H|^2} \geq \frac{q^{2r}}{r^2q^2} = \frac{q^{2r-2}}{r^2} \geq \frac{8^{2r-2}}{r^2}.$$

Elementary calculations show that $8^{2r-2} \geq r^2$ for all natural numbers r. Therefore, $|M/C_M(A)| \geq |A|^2$, as desired.

Case 2. *The simple direct factors of H have the form* $PSL(2, 3^{2n+1})$. Here, let $q = 3^{2n+1}$. By (1.15), (1.17), and Lemma II.2.7, $G \cong PSL(2, q)$. It is easy to see that $|G|_2 = 4$ and that

(1.21) all involutions in G are conjugate

and

(1.22) for each involution $x \in G$, $C(x)$ is dihedral of order $q + 1$.

Now take $y \in A^{\#}$. Choose $x \in A - \langle y \rangle$ if $|A| = 4$; otherwise choose any involution x of $C(y) - \langle y \rangle$.

Since $q = 3^{2n+1} = 9^n \cdot 3$, it follows that

(1.23) $q + 1 \equiv 3 + 1 \equiv 4 \pmod 8$, $q + 1 \not\equiv 0 \pmod 3$,   and

$$q + 1 \equiv (-1)^n \cdot 3 + 1 \not\equiv 0 \pmod 5.$$

As $n \geq 1$, $q + 1 \geq 28$. Therefore, $(q + 1)/4$ is an odd integer greater than one. Let r be the smallest prime divisor of $(q + 1)/4$. By (1.23), $r \geq 7$. By (1.22) and our choice of x, it follows that y inverts some subgroup R of order r in $C(x)$.

Now, $\langle x, y, R \rangle = \langle x \rangle \times \langle y, R \rangle$. By Lemma 1.1, $\langle R, y \rangle$ acts faithfully on $C_M(x)$. Let $M^* = C_M(x)$. By Lemma 1.2(c),

$$(1.24) \qquad\qquad |M^*/C_{M^*}(y)| \geq 2^3.$$

If $A = \langle y \rangle$, then $|A| = 2$ and

$$|A|^2 = 2^2 < 2^3 \leq |M^*/C_{M^*}(y)| \leq |M/C_M(y)| = |M/C_M(A)|,$$

as desired. Assume $A \neq \langle y \rangle$. Then $|A| = 4$ and, by our choice of x, $x \in A$. By (1.24),

$$|M/C_M(x)| = |M/C_M(y)| \geq |M^*/C_{M^*}(y)| \geq 2^3.$$

Therefore,

$$|M/C_M(A)| = |M/C_M(x)||C_M(x)/C_M(A)|$$

$$= |M/C_M(x)||M^*/C_{M^*}(y)| \geq 2^6 > 2^4 = |A|^2,$$

as desired. This completes the proof of Case 2 and of Theorem A.

Appendix A2. Corrections and Additions to GL

In this section, we list nontrivial changes to be made in GL. We preface each change by its page and line number; x, y denotes line y of page x, while x, $-y$ denotes the yth line from the bottom of page x.

1, footnote	We are indebted to N. Blackburn and W. Specht
7, -4	(or on $\delta(S)$)
8, -12	(1968). A similar proof of Lemma 4.2 is given in pp. 412–417
20, 10	$N_{\underset{G}{-}}(\bar{S}) = \bar{N}$
23, -16 and -15	of a p-stable group given in (Gorenstein, 1968).
24, 14	$G = C(Q^p)N(C_S(Q^p))$.
28, -3 and -2	Choose G^* and S^* to satisfy the hypothesis of (8.2) and violate its conclusion. Take X, Y, and g as in (8.8)
35, -6	(c) if $C(T) \subseteq T$
41, -3	Professor Gorenstein has used (14.1) in his lectures.
42, 1	proved in pp. 271–274
42, 4	(b) If $Q \subseteq P$, $A \in \mathfrak{A}(P)$, and $A \subseteq Q$, then $A \in \mathfrak{A}(Q)$ and
44, 14	is that whenever $T \subseteq S$, then $N(T)/C(T)$ have a normal Sylow 2-subgroup.
45, -18 and -17	Case (c) may be simplified by the methods of Appendix A.1 (of GL).
45, -2 and -1	joint result (1968) of Thompson and the author:
48, 7	results follow from Lemma 4.5 (of GL) and Theorem III.3.1 (of this work)
49, -2	of Glauberman (1968c)
52, -4	for $v \in V_0 \cap Z(V)$
59, 16 and 17	whenever A and B are nonempty subsets of S, $h \in G$, and $A^h = B$, then there exists $g \in G$ such that gh^{-1} centralizes A and A is $\mathfrak{F}^*$-conjugate to B via g.
60, -12	Goldschmidt (1970)
60, -9	generalizes Theorem 7.9
61, 6	Glauberman, 1968 c)
pp. 63–64	The papers of Shult and Thompson without journal citations are special cases of Theorem III.1 and [42] of this work and will not be published. The other references without journal citations will be found in the bibliography of this work.

Appendix A3.

In this section, we feature my review of Bernd Stellmacher's article, *A characteristic subgroup* Σ_4-*free groups*, Israel J. Math 94 (1996), 367–379. the review originally appeared in Mathematical Reviews, April, 1997, as review 97d:20018.

This paper is a remarkable advance in an area of finite group theory that began about forty years ago. In 1959, John Thompson proved his celebrated theorem that the Frobenius kernel of a Frobenius group is nilpotent. Five years later, he introduced a characteristic subgroup $J(S)$ of a p-group S and used it to give a much simpler proof, which depended partly on analyzing the structure of a special type of solvable group [J. Algebra 1 (1964), 43–46; MR 29 #4793]. Shortly afterwards, he extended this analysis to obtain a factorization of a solvable (or more generally, p-solvable) group H as a product of two subgroups depending on a Sylow p-subgroup of H [Pacific J. Math. 16 (1966), 371–372; MR 32 #5735].

Each of these results reduces to a case in which there are a prime p and a p-solvable group H such that p divides the order of each non-trivial normal subgroup of H. Let S be a Sylow p-subgroup of H and $Z(S)$ be the center of S. Thompson defined $J(S)$ to be the subgroup of S generated by the abelian subgroups of maximal rank in S. For his factorization result, he proved that $H = C_H(Z(S))N_H(J(S))$ unless $p \leq 3$ and the group $\mathrm{SL}(2, p)$ is involved in H, i.e., there exist subgroups K, L of H such that $L \lhd K$ and $K/L \cong \mathrm{SL}(2, p)$. (He also obtained a stronger "triple factorization" when $p \geq 5$.) The conclusion may be false if $\mathrm{SL}(2, p)$ is involved. For instance, for $p = 2$ and $H = \Sigma_4$ (the symmetric group of degree 4), $C_H(Z(S)) = N_H(J(S)) = S$, so that $C_H(Z(S))N_H(J(S)) < H$.

Now let p be a fixed prime and S be a fixed non-trivial p-group. For any group H, there is a unique maximal normal p-subgroup, called $O_p(H)$. By extending Thompson's methods, the reviewer showed in 1967 (for a slightly different subgroup $J(S)$) that, for p odd, $Z(J(S)) \lhd H$ in the situation above or, more generally, whenever (A) S is a Sylow p-subgroup of H, and $O_p(H)$ contains its centralizer $C_H(O_p(H))$; and (B) $\mathrm{SL}(2, p)$ is not involved in H [Canad. J. Math. 20 (1968), 1101–1135; MR 37 #6365]. Thus, there is a non-trivial characteristic subgroup of S that is normal (in fact, characteristic) in H, and independent of the choice of H. This result was applied to the study of a simple non-abelian group G in which S is a Sylow p-subgroup; then $N_G(Z(J(S)))$ contains every subgroup H of G satisfying (A) and (B). Later, I. M. Isaacs found a simpler proof by using a subgroup defined by L. Puig instead of $Z(J(S))$ (Appendix B in *Local analysis for the odd order theorem* by H. Bender and the reviewer [Cambridge Univ. Press, Cambridge, 1994; MR 96h:20036]). More recently, the author used the "amalgam method" to prove the existence of an analogous subgroup without having to specify the subgroup [Proc. Amer. Math. Soc. 109 (1990), no. 4, 925–929; MR 90k:20045; errata, ibid. 114 (1992), no. 2, 588; MR 92e:20012].

These results naturally raise the question of an analogue for $p = 2$, i.e., a non-trivial characteristic subgroup $W(S)$ of S such that $W(S) \lhd H$ for every H satisfying (A) and (B). (Examples show that $Z(J(S))$ does not work.) But this case is much more difficult. For odd p, hypothesis (B) yields a strong condition (p-stability) about commutators for normal subgroups. But, for $p = 2$, it appears to yield comparatively weak information. Moreover, in many ways, 2-groups are much more complicated than p-groups for odd p. Therefore, this problem stimulated considerable effort for nearly thirty years. In the present paper, the author finally solves the problem by showing that $W(S)$ exists.

A special case of this result for a family of solvable groups was proved by M. Hayashi [Math. Z. 181 (1982), no. 2, 215–227; MR 84f:20019]. A weaker consequence of (A) and (B) for $p = 2$, namely, a "triple factorization" similar to Thompson's above and inspired by a special case in Thompson's N-group work, was proved by the reviewer in 1977 [*Factorizations in local subgroups of finite groups*, Amer. Math. Soc., Providence, R.I., 1977; MR 57 #9839 (pp. 27–28)].

As in the case of Thompson's factorization, the group Σ_4 violates the hypothesis and the conclusion of the author's theorem. Moreover, it can be shown that for $p = 2$, conditions (A) and (B) are equivalent to the conditions (I) H is Σ_4-free (i.e., Σ_4 is not involved in H) and (II) S is a Sylow 2-subgroup of H and $C_H(O_2(H)) \leq O_2(H)$. Thus, Σ_4 is the "root" of all counterexamples.

The main part of the paper proves the conclusion under the assumptions (I), (II), and an additional condition: (III) Every non-abelian simple section of H is isomorphic to $\mathrm{Sz}(2^m)$ or $\mathrm{PSL}_2(3^m)$ for some odd m. A similar approach was followed in the reviewer's triple factorization paper. As in that paper [op. cit., p. 29], one easy consequence of the main part and of a theorem of D. Goldschmidt (about groups containing a strongly closed abelian 2-subgroup) is the classification of all Σ_4-free non-abelian simple groups. (However, this is not mentioned by the author.) This, in turn, shows that the only Σ_3-free non-abelian groups are the groups in (III). Thus, since $\mathrm{SL}(2,2) \cong \Sigma_3$, condition (III) follows from (B) and is finally seen to be superfluous. A further consequence is that the Suzuki groups are the only non-abelian simple groups of order not divisible by 3.

The proof is largely inspired by the "amalgam method", although it does not explicitly use it. For example, to define $W(S)$, the author embeds a certain family of groups H as above for which $J(S) \triangleleft H$ into one huge group G_1 (their amalgamated product over S), which may be infinite. He then defines $\Omega_1(Z(S))$ to be the group of all $x \in Z(S)$ for which $x^2 = 1$ (as usual) and $W(S)$ to be the group generated by all conjugates of $\Omega_1(Z(S))$ in G_1. Then $W(S)$ is a characteristic subgroup of S contained in $Z(J(S))$, and he must prove that $W(S) \triangleleft H$ for every group H satisfying (I), (II), and (III), regardless of whether $J(S) \triangleleft H$.

The main preliminary results about a group H as above are given in Section 2. By exploiting (III) and properties of $\mathrm{Sz}(2^m)$ and $\mathrm{PSL}_2(3^m)$, the author finds and analyzes a useful subgroup L of H such that $L/O_2(L)$ is dihedral or a Suzuki group. Most of Section 2 is devoted to studying the action of H or of L on a GF(2)-module V. An essential tool in earlier work was that $|V/C_V(A)| > |A/C_A(V)|$ if $O_2(H/C_H(V)) = 1$, A is a subgroup of S, and $A/C_A(V)$ is a non-trivial elementary abelian 2-group. But this did not cover some important situations in which $O_2(H/C_H(V)) > 1$. Here, the author extends this inequality to these situations by using an earlier result, (2.3), and a remarkable new result, (2.9).

The proof proper appears in the last section (Section 3), and can only be described as a tour de force. The author assumes the existence of a group H in which $W(S)$ is not normal. By using his information about the amalgam construction, the subgroup L, and GF(2)-modules to their fullest extent, he eventually obtains some delicate inequalities that lead to a contradiction. Somewhat surprisingly, the case in which H is solvable is much more difficult than the non-solvable case.

{Reviewer's remarks: The author neglected to give a choice of $J(S)$. One may take Thompson's definition above or the subgroup generated by the elementary abelian subgroups of maximal rank in S. The author has a correction for page 378: Replace "≤ 2, and again 2.6" by "≤ 4 and $|B/(B \cap O_2(U))| = |B/(B \cap X_1)| = 2$, and again 2.6 applied to L and X/X_1".}

Bibliography

1. J. Alperin, *Finite groups viewed locally*, Bull. Amer. Math. Soc. 83 (1977), 1271–1285.

2. J. Alperin and R. Lyons, *On conjugacy classes of p-elements*, J. Algebra 19 (1971), 536–537. MR 44 #6810.

3. E. Artin, *The orders of the linear groups*, Comm. Pure Appl. Math. 8 (1955), 355–365. MR 17, 12.

4. M. Aschbacher, *A characterization of Chevalley groups over fields of odd order*, Ann. of Math. (2) 106 (1977), 353–468; errata, ibid. (2) 111 (1980), 411–414, MR 58 #16865a, b, 81j:20026.

5. ——, *A pushing up theorem for characteristic 2 type groups*, Illinois J. Math. 22 (1978), 108-120. MR 57 #9830.

6. ——, *On the failure of the Thompson factorization in 2-constrained groups*, Proc. London Math. Soc. (3) 43 (1981), 425–449. MR 83a:20019a.

7. ——, *A factorization theorem for 2-constrained groups*, Proc. London Math. Soc. (3) 43 (1981), 450–477. MR 83a:20019b.

8. ——, *Thin finite simple groups*, J. Algebra 45 (1978), 50–152. MR 82j:20032

9. B. Baumann, *Endliche nichtauflösbare Gruppen mit einer nilpotenten maximalen Untergruppen*, J. Algebra 38 (1976), 119–135.

10. ——, *Über endliche Gruppen mit einer zu $L_2(2^n)$ isomorphen Faktorgruppe*, Proc. A. M. S. 74 (1979), no. 2, 215–222, MR 80g: 20024.

11. H. Bender, *On the uniqueness theorem*, Illinois J. Math. 14 (1970), 376–384. MR 41 #6959.

12. ——, *Goldschmidt's 2-signalizer functor theorem*, Israel J. Math. 22 (1975), 208–213. MR 52 #10882.

13. A. Borel and J. Tits, *Éléments unipotents et sous-groupes paraboliques de groupes reductifs*. I, Invent. Math. 12 (1971), 95–104. MR 45 #3419.

14. H. Cartan and S. Eilenberg, *Homological algebra*, Princeton Univ. Press, Princeton, N. J., 1956. MR 17, 1040.

15. R. Carter, *Simple groups of Lie type*, Wiley, New York, 1972.

16. U. Dempwolff, *On primitive permutation groups whose stabilizer of a point induces $L_2(q)$ on a suborbit*, Illinois J. Math. 20 (1976), 48–64; correction in Illinois J. Math. 21 (1977), 427.

17. L. Dornhoff, *Group representation theory*, Part A, Marcel Dekker, New York, 1971. MR 50 #458A.

18. W. Feit, *The current situation in the theory of finite simple groups*, Proc. Internat. Congress Math. (1970), Gauthier-Villars, Paris, 1971.

19. W. Feit and J. G. Thompson, *Solvability of groups of odd order*, Pacific J. Math. 13 (1963), 775–1029. MR 29 #3538.

20. P. Ferguson, *An application of the Glauberman-Goldschmidt Theorem to $3'$-homogeneous groups*, J. Algebra 43 (1976), 212–215.

21. D. Finkel, *Local control and factorization of the focal subgroup*, Pacific J. Math. 45 (1973), 113–128. MR 47 #5106.

22. L. R. Fletcher, B. Stellmacher, and W. B. Stewart, *Endliche Gruppen, die kein Element der Ordnung 6 enthalten*, Quarterly J. Math. (Oxford) (2) 28 (1977), 143–154.

23. T. M. Gagen, *Topics in finite groups*, Cambridge Univ. Press, London and New York, 1976.

24. G. Glauberman, *Central elements of core-free groups*, J. Algebra 4 (1966), 403–420. MR 34 #2681.

25. ——, *A characteristic subgroup of a p-stable group*, Canad. J. Math. 20 (1968), 1101–1135. MR 37 #6365.

26. ——, *Isomorphic subgroups of finite p-groups*. I, II, Canad. J. Math. 23 (1971), 983–1022, 1023–1039.

27. ——, *A sufficient condition for p-stability*, Proc. London Math. Soc. (3) 25 (1972), 253–287.

28. ——, *Direct factors of Sylow 2-subgroups*. I, II, J. Algebra 28 (1974), 133–161, 162–173. MR 49 #9076.

29. ——, *Factorizations for 2-constrained groups*, Proc. London Math. Soc. (3) 41 (1980), 386–438. MR 82f:20041.

30. ——, *A characterization of the Goldschmidt groups* (unpublished, superseded by this monograph).

31. D. Goldschmidt, *On the 2-exponent of a finite group*, Ph. D. Thesis, University of Chicago, Chicago, Illinois, 1969.

32. ——, *A conjugation family for finite groups*, J. Algebra 16 (1970), 138–142. MR 41 #5489.

33. ——, *An application of Brauer's second main theorem*, J. Algebra 20 (1972), 72–77. MR 45 #364.

34. ——, *2-Fusion in finite groups*, Ann. of Math. (2) 99 (1974), 70–117. MR 49 #407.

35. ——, *Strongly closed 2-subgroups of finite groups*, Ann. of Math. 102 (1975), 475–489.

36. ——, unpublished.

37. D. Gorenstein, *Finite groups*, Harper and Row, New York, 1968. MR 38 #229.

38. ——, *Finite simple groups and their classification*, Israel J. Math. 19 (1974), 5–66. MR 50 #13244.

39. M. Hall, *The theory of groups*, Macmillan, New York, 1959. MR 21 #1996.

40. M. Hayashi, *A remark on G. Glauberman's theorem*, J. Math. Soc. Japan 27 (1975), 256–257. MR 53 #605.

41. G. Higman, *Odd characterizations of finite simple groups*, Lecture Notes, University of Michigan, Ann Arbor, Michigan, 1968.

42. C.-Y. Ho, *Chevalley groups of odd characteristic as quadratic pairs*, J. Algebra 41 (1976), 202–211.

43. ——, *Quadratic pairs for odd primes*, Bull. Amer. Math. Soc. 82 (1976), 941–943.

44. B. Huppert, *Endliche Gruppen.* I, Springer-Verlag, Berlin and New York, 1967. MR 37 #302.

45. Z. Janko, *A new finite simple group with abelian Sylow 2-subgroups and its characterization*, J. Algebra 3 (1966), 147–186. MR 33 #1359.

46. Z. Janko and J. G. Thompson, *On a class of finite simple groups of Ree*, J. Algebra 4 (1966), 274–292. MR 34 #1386.

47. W. Knapp, *Primitive Permutations gruppen mit einem zweifach primitiven Subkonstituenten*, J. Algebra 38 (1976), 146–162.

48. R. Niles, *Pushing-up in finite groups*, J. Algebra 57 (1979), 26–63. MR 80g:20027.

49. M. Pettet, *A note on finite groups having a fixed-point-free automorphism*, Proc. Amer. Math. Soc. 52 (1975), 79–80.

50. ——, *On a theorem of Goldschmidt applied to groups with a coprime automorphism*, Canad. J. Math. 28 (1976), 201–206.

51. M. B. Powell and G. Higman (Editors), *Finite simple groups*, Academic Press, New York, 1971. MR 48 #6228.

52. L. Puig, *Structure locale dans les groupes finis*, Bull. Math. Soc. France, Mémoire 47 (1976).

53. L. Schiefelbusch, *On the transfer homomorphism*, Comm. Algebra 3 (1975), 295–317. MR 51 #8235.

54. ——, *Transfer and weakly closed subgroups*, Comm. Algebra 7 (1979), no. 4, 333–340. MR 80e: 20031.

55. E. Shult, *Some analogues of Glauberman's Z^*-theorem*, Proc. Amer. Math. Soc. 17 (1966), 1186–1190. MR 33 #5736.

56. C. C. Sims, *Graphs and finite permutation groups*, Math. Z. 95 (1967), 76–86. MR 34 #4348.

57. M. Suzuki, *On a class of doubly transitive permutation groups*, Ann. of Math. (2) 75 (1962), 105–145. MR 25 #112.

58. J. G. Thompson, *Factorizations of p-solvable groups*, Pacific J. Math. 16 (1966), 371–372. MR 32 #5735.

59. ——, *Nonsolvable finite groups all of whose local subgroups are solvable.* I, II, III, IV, V, VI, Bull. Amer. Math. Soc. 74 (1968), 383–437; ibid., Pacific J. Math. 33 (1970), 451–536; ibid., Pacific J. Math. 39 (1971), 483–534; ibid., Pacific J. Math. 48 (1973), 511–592; ibid., Pacific J. Math. 50 (1974), 215–297; ibid., Pacific J. Math. 51 (1974), 573–630. MR 37 #6367; 43 #2072; 47 #1933; 51 #5745.

60. ——, *Quadratic pairs* (unpublished).

61. ——, *Simple groups of order prime to* 3. I (unpublished).

62. J. H. Walter, *Finite groups with abelian Sylow 2-subgroups of order* 8, Invent. Math. 2 (1967), 332–376. MR 36 #1531.

63. H. N. Ward, *On Ree's series of simple groups*, Trans. Amer. Math. Soc. 121 (1966), 62–89. MR 33 #5752.

64. H. Wielandt, *p-Sylowgruppen und p-Faktorgruppen*, J. Reine Angew. Math. 182 (1940), 180–193. MR 2, 216.

65. ———, *Subnormal subgroups and permutation groups*, Lecture Notes, Ohio State University, Columbus, Ohio, 1971.

66. ———, *Kritierien für Subnormalität in endlichen Gruppen*, Math. Z. 138 (1974), 199–203. MR 50 #10072.

67. W. J. Wong, *Determination of a class of primitive permutation groups*, Math. Z. 99 (1967), 235–246. MR 35 #5502.

68. T. Yoshida, *Character-theoretic transfer*, J. Algebra 52 (1978), 1–38. MR 58 #11095

69. K. Harada, *On Yoshida's transfer*, Osaka Math. J. 15 (1978), no. 3, 637–646. MR 80c: 20017.

References added in Second printing

70. P. J. Cameron, C. E. Praeger, J. Saxl, and G. M. Seitz, *On the Sims Conjecture and distance transitive groups*, Bull. London Math. Soc. 15 (1983), 499–506. MR 85g:20006.
71. G. Glauberman, *Large abelian subgroups of finite p-groups*, J. Algebra 196 (1997), 301–338. MR 98h:20029.
72. G. Glauberman and R. Niles, *A pair of characteristic subgroups for pushing-up in finite groups*, Proc. London Math.Soc. (3) 46 (1983), 411–453. MR 84g:20042.
73. D. Gorenstein, *Finite Simple Groups: An Introduction to their Classification*, Plenum Press, New York, 1982. MR 84j:20002.
74. D. Gorenstein, R. Lyons, R. Solomon, *The Classification of the Finite Simple Groups*, no. 3, American Mathematical Society, Providence, 1998. MR 98j:20011.
75. D. Holt, *More on the local control of Schur multipliers*, Quart. J. Math. Oxford Ser. (2) 31 (1980), 191–208. MR 81j:20022.
76. M. Miyamoto, *An affirmative answer to Glauberman's conjecture*, Pacific J. Math. 102 (1982), 89–105. MR 84b:20060.
77. A. A. Premet and I. D. Suprunenko, *Quadratic modules for Chevalley groups over fields of odd characteristics*, Math. Nachr. 110 (1983), 65–96. MR 85c:20037.

Index

Alperin, J., 1–2, 5–7, 10–11
Aschbacher, M., 54–56, 58

Baer, R., 5
Baumann, B., 54, 57
Bender, H., 56
Burnside, W., 1, 4, 7

chief factor (central, noncentral, with-
    in a subgroup), 3
cohomology group, 57
commutator, 19
conjugacy functor, 7–11, 26
conjugacy of sequences, 6
conjugation family, 6
containment of fusion, 6–8, 26
control of strong fusion, 2, 10–11, 13,
    15, 18, 26, 45, 49–51, 57–58
    counterexamples, 26, 53–54
control of transfer, 2, 10–11, 15, 20,
    26, 57
counterexamples, 26, 53–54

Dedekind, R., 13
Dempwolff, U., 54
Dickson, L. E., 33, 46
Dolan, S., 51, 57

E-group, 29

factorizations, 18, 27–28, 48–49
$\mathcal{F}$-conjugacy, 5–6
Feit, W., 19
Ferguson, P., 51
Finkel, D., 7, 26
fixed-point-free groups of automor-
    phisms, ix, 50

Fletcher, L. R., 57
Focal Subgroup Theorem, 6, 56
$F(p)$, 48–50
Frattini argument, 4
Frattini subgroup, 3–5
fusion, 1–2

Glauberman, G., 27, 56
global structure, ix, 55–59
GL, summary, vii
Goldschmidt, D., 45–46, 54, 56–57
Goldschmidt group, 29, 46
Gorenstein, D., 2, 7, 10–11, 28, 56–57
group satisfying various restrictions
    $F(p)$ not involved, 50
    no elements of order six, 57
    p-stable, 3, 27, 29, 49
    $Qd(p)$ not involved, 49–50, 52–53
    Ree type, 46
    restricted for several primes, ix
    S^3 not involved, 28, 30–31, 48, 51
    S^4-free, 27, 29, 31, 45, 47–49
    Sylow 2-subgroup maximal, 57
    Sylow 2-subgroup normal, 50–51

Harada, K., 22, 58
Harris, M., 56–57
Hayashi, M., 50
Higman, G., 57
Ho, C.-Y., 55

inversion of elements, 3
involution, 3
involvement, 5
Isaacs, M., 22

75